KT-392-289

DR005569
CORNWALL COLLEGE

Conserving Plant Genetic Diversity in Protected Areas

Population Management of Crop Wild Relatives

Conserving Plant Genetic Diversity in Protected Areas

Population Management of Crop Wild Relatives

Edited by

José María Iriondo

Area de Biodiversidad y Conservación
ESCET
Universidad Rey Juan Carlos
Madrid, Spain

Nigel Maxted

School of Biosciences
University of Birmingham
Birmingham, UK

Mohammad Ehsan Dulloo

Bioversity International
Rome, Italy

CABI is a trading name of CAB International

CABI Head Office
Nosworthy Way
Wallingford
Oxfordshire OX10 8DE
UK

Tel: +44 (0)1491 832111
Fax: +44 (0)1491 833508
E-mail: cabi@cabi.org
Website: www.cabi.org

CABI North American Office
875 Massachusetts Avenue
7th Floor
Cambridge, MA 02139
USA

Tel: +1 617 395 4056
Fax: +1 617 354 6875
E-mail: cabi-nao@cabi.org

A catalogue record for this book is available from the British Library, London, UK.

Library of Congress Cataloging-in-Publication Data

Conserving plant genetic diversity in protected areas: population management of crop wild relatives / editors: José M. Iriondo, Nigel Maxted and M. Ehsan Dulloo.
p. cm.
Includes bibliographical references and index.
ISBN 978-1-84593-282-4 (alk. paper)
1. Germplasm resources, Plant. 2. Crops--Germplasm resources. 3. Genetic resources conservation. 4. Plant diversity conservation. I. Iriondo, José M. II. Maxted, Nigel. III. Dulloo, M. Ehsan (Mohammad Ehsan) IV. Title.

SB123.3.C666 2008
639.9'9--dc22 2007039904

ISBN: 978 1 84593 282 4

Typeset by SPi, Pondicherry, India
Printed and bound in the UK by Biddles Ltd, King's Lynn

Contents

Preface

This book is about the conservation of genetic diversity of wild plants *in situ* in their natural surroundings, primarily in existing protected areas but also outside conventional protected areas. A lot of effort has been dedicated to conserving plant biodiversity, but most of this has focused on rare plant communities or individual species threatened with extinction. Similarly, while much has been done to collect and conserve crop genetic diversity *ex situ* in gene banks, very little consideration has been given to conserving intraspecific genetic diversity *in situ* and in particular while designing protected areas.

Why should we care about the genetic aspect of biodiversity conservation? Genetic diversity is in fact essential for any species to underwrite its ability to adapt and survive in the face of environmental change. After all, the history of life is a history of change, a constant adaptation of life forms to a dynamic world. However, the rate at which our planet's environment is now changing is dramatically increasing due to the activities of humans around the world. Therefore, the relevance of the genetic diversity of plants and other life forms to adapt to these changing conditions is now higher than ever. Furthermore, as humans we also face the uncertainty of our actions in the future. In an environmentally dynamic world with a constantly increasing population and limited resources, we need to conserve genetic diversity for our own food and environmental security.

Throughout the last 10,000 years, farmers have cultivated plants of approximately 10,000 species to provide food, medicines and shelter, and through careful breeding have generated an extraordinary diversity of crops adapted to the local characteristics of each site. In the last century, our intimate knowledge of the genetic basis of inheritance sparked a revolution in agriculture that resulted in a quantum leap in production but these high-yielding varieties tended to be genetically uniform. As farmers have progressively abandoned their traditional varieties and landraces and shifted to the cultivation of more productive modern cultivars, the number of food crops and their genetic diversity has dangerously narrowed. Today, over 50% of food production from plant origin is derived from only three crop species and 90% comes from the first 25 crops. This situation, coupled with high levels of genetic erosion in these crops through the abandonment of traditional genetically diverse landrace varieties, has placed food production in a very vulnerable situation with regard to future changes in physical environmental conditions and the arrival of new races of pests and pathogens. Many countries and the international community have been aware of this problem and during the past few decades have consequently established germplasm banks to store the genetic diversity contained in the vanishing traditional varieties and landraces.

More recently, attention has been brought to conserving the genetic diversity present within wild plants, particularly those closely related to crop species, known as crop wild relatives (CWR). The much needed genes that could provide the required adaptation to changing environmental conditions and tolerance or resistance to new strains of pests and pathogens are probably already present in CWR and can be easily transferred when needed. Conservation in germplasm banks is an effective way of preserving large amounts of crop germplasm that may be used for future plant breeding. Nevertheless, a major drawback of this methodology is that the genetic evolution of this germplasm is 'frozen' because the germplasm is maintained in a latent life form (i.e. seeds). Also, the costs of location and sampling the genetic diversity of all wild plants would be too prohibitive. Furthermore, *in situ* conservation necessarily involves the protection of habitat and ecosystems, so engendering broader ecological integrity and resultant human well-being – after all, making genes available to breeders is an important, but only one, use of biodiversity.

Today there is a consensus among the conservation community that the best way of conserving a species and its genetic diversity is *in situ*, i.e. through the conservation of their populations in their natural habitats. In this way, generation after generation, natural populations can evolve and adapt to physical environmental trends and to changes in the web of interactions with other life forms. Nevertheless, conservation always comes at a cost and the land that is set aside for *in situ* conservation may not be compatible with some human activities. Therefore, any conservation strategy must always keep in mind the socio-economic environment and the scale of values, and the interests that human society has at each location.

Wild plant species are fundamental constituents of all kinds of habitats and ecosystems. Although many occur in natural ecosystems and pristine habitats (whether protected or not), others, particularly the close CWR of our major crops, are present in perturbed habitats and human-transformed habitats such as those linked to agriculture or transport infrastructures. In this book we focus on the establishment and management of genetic reserves for conserving plant genetic diversity in protected areas. There are several advantages for this. The first one is the economic savings in infrastructure and maintenance when the genetic reserve is located in an existing protected area, as well as the lack of problems related to setting aside a territory that may be of interest for human development activities. There is in fact a mutual benefit in the establishment of a genetic reserve in a protected area. Genetic reserves for CWR are likely to be welcomed by protected area managers since their establishment will undoubtedly increase the perceived natural assets and values of the site. The second advantage relates to the long-term sustainability of the genetic reserve. If the genetic reserve is not in a protected area, there is no guarantee that the land will be kept as a reserve in the long term due to shifting political and socio-economic decisions.

Although the focus of this book is the *in situ* conservation of the genetic diversity of species related to crops, there is essentially no fundamental conservation distinction between those wild species closely related to crops and those that are not. Perhaps the only difference is the potential use of the diversity once it is conserved. The principles outlined in what follows are equally applicable for the *in situ* genetic conservation of any wild plant species, whether the aim is to maintain a species threatened by habitat fragmentation, over-collection from the wild or a species that has potential use as a gene donor to our crops.

This book is arranged in a logical, sequential structure to help guide the conservationists in the establishment of a reserve for the conservation and management of genetic diversity of wild plant species. After an introductory chapter where the main concepts are presented, the selection of the genetic reserve location and its design are discussed in Chapter 2. Next, Chapter 3 presents the management plan that must be inherent to any *in situ* conservation strategy in a genetic reserve and Chapter 4 describes the monitoring activities that are required for the long-term maintenance of wild populations. However, the target populations in genetic reserves may not always be in an optimum state and, consequently, a set of restorative actions on the target population and/or the surrounding habitat may be needed. Thus, Chapter 5 shows the main population and habitat recovery techniques that are currently available. We have already stated that one of the final goals of CWR conservation in reserves is to provide a wealth of genetic diversity that may be used by plant breeders to respond to future challenges in food production. In order to make this possible and to maximize the benefits of this initiative, Chapter 6 explores the safety and utilization linkages of genetic reserves with germplasm banks and other plant genetic resource repositories to facilitate a flux of germplasm and related information that may be used by plant breeders. Finally, Chapter 7 provides an economic assessment of genetic reserves along with some policy considerations and presents some of the challenges and trends that we perceive for the future.

Obviously, the *in situ* conservation of wild plant genetic diversity should not be restrained to protected areas alone, especially as some species are often associated with human-moderated ecosystems. Many of the indications provided in this book can readily be applied in initiatives dealing with the conservation of wild plant genetic diversity in environments outside formal protected area networks. Nevertheless, this is one of the issues that should be studied in more detail in future activities in CWR conservation.

José María Iriondo
Nigel Maxted
Mohammad Ehsan Dulloo

June 2007

Contributors

A. Asdal, *Norwegian Genetic Resources Centre, PO Box 115, N-1431 Aas, Norway. E-mail: asmund.asdal@skogoglandskap.no; Fax: + 47 37 044 278.*

L. de Hond, *Area de Biodiversidad y Conservación, ESCET, Universidad Rey Juan Carlos, c/Tulipán s/n, E-28933 Móstoles, Madrid, Spain. E-mail: optima-madrid@telefonica.net; Fax: + 34 916 647 490.*

M.E. Dulloo, *Bioversity International, Via dei Tre Denari 472/a, 00057 Maccarese, Rome, Italy. E-mail: e.dulloo@cgiar.org; Fax: + 39 0 661 979 661.*

J.M.M. Engels, *Bioversity International, Via dei Tre Denari 472/a, 00057 Maccarese, Rome, Italy. E-mail: j.engels@cgiar.org; Fax: + 39 0 661 979 661.*

B. Ford-Lloyd, *School of Biosciences, University of Birmingham, Edgbaston, Birmingham B15 2TT, UK. E-mail: b.ford-lloyd@bham.ac.uk; Fax: + 44 121 414 5925.*

L. Guarino, *Global Crop Diversity Trust, c/o FAO, Viale delle Terme di Caracalla, 00153 Rome, Italy. E-mail: luigi.guarino@croptrust.org; Fax: + 39 06 570 54951.*

J.M. Iriondo, *Area de Biodiversidad y Conservación, ESCET, Universidad Rey Juan Carlos, c/Tulipán s/n, E-28933 Móstoles, Madrid, Spain. E-mail: jose.iriondo@urjc.es; Fax: + 34 916 647 490.*

A. Jarvis, *Bioversity International and International Centre for Tropical Agriculture, c/o CIAT, Apartado Aereo 6713, Cali, Colombia. E-mail: a.jarvis@cgiar.org; Fax: + 57 24 450 096.*

S.P. Kell, *School of Biosciences, University of Birmingham, Edgbaston, Birmingham B15 2TT, UK. E-mail: s.p.kell@bham.ac.uk; Fax: + 44 121 414 5925.*

H. Korpelainen, *Department of Applied Biology, PO Box 27 (Latokartanonkaari 7), FIN-00014 University of Helsinki, Finland. E-mail: helena.korpelainen@helsinki.fi; Fax: + 358 919 158 727.*

J. Labokas, *Institute of Botany, Žaliųjų Ežerų g. 49, LT-08406 Vilnius, Lithuania. E-mail: juozas.labokas@botanika.lt; Fax: + 370 52 729 950.*

E. Laguna, *Centro para la Investigación y Experimentación Forestal (CIEF), Generalitat Valenciana. Avda. País Valencià, 114, E-46930 Quart de Poblet, Valencia, Spain. E-mail: laguna_emi@gva.es; Fax: + 34 961 920 258.*

A. Lane, *Bioversity International, Via dei Tre Denari 472/a, 00057 Maccarese, Rome, Italy. E-mail: a.lane@cgiar.org; Fax: + 39 0 661 979 661.*

F. Lefèvre, *INRA, URFM, Unité de Recherches Forestières Méditerranéennes (UR629) Domaine Saint Paul, Site Agroparc, F-84914 Avignon Cedex 9, France. E-mail: lefevre@avignon.inra.fr; Fax: + 33 432 722 902.*

L. Maggioni, *Bioversity International, Via dei Tre Denari 472/a, 00057 Maccarese, Rome, Italy. E-mail: l.maggioni@cgiar.org; Fax: + 39 0 661 979 661.*

N. Maxted, *School of Biosciences, University of Birmingham, Edgbaston, Birmingham, B15 2TT, UK. E-mail: n.maxted@bham.ac.uk; Fax: + 44 121 414 5925.*

Acknowledgements

This volume grew out of the EC-funded project, PGR Forum (the European crop wild relative diversity assessment and conservation forum – EVK2-2001-00192 – http://www.pgrforum.org/). As such, many of the concepts presented in this volume were stimulated by PGR Forum discussions. PGR Forum was funded by the EC Fifth Framework Programme for Energy, Environment and Sustainable Development and the editors wish to acknowledge the support of the European Community in providing the forum for discussion and publication of this volume.

1 Introduction: The Integration of PGR Conservation with Protected Area Management

N. Maxted,[1] J.M. Iriondo,[2] M.E. Dulloo[3] and A. Lane[3]

[1]School of Biosciences, University of Birmingham, Edgbaston, Birmingham, UK; [2]Área de Biodiversidad y Conservación, Depto. Biología y Geología, ESCET, Universidad Rey Juan Carlos, Madrid, Spain; [3]Bioversity International, Rome, Italy

1.1 Plant Conservation, Plant Genetic Resources and *In Situ* Conservation

The Convention on Biological Diversity (CBD, 1992) fundamentally changed the practice of plant conservation by placing greater emphasis on the *in situ* conservation of biological diversity, that is the natural diversity of ecosystems, species and genetic variation, and employing *ex situ* conservation as a safety back-up action to preferred *in situ* activities. The Convention also stressed the direct link between conservation and use, and the requirement for fair and equitable sharing of benefits between the original resource managers and those responsible for its exploitation. Certainly in the context of socio-economically important plant species conservation this was a distinct change switching the emphasis away from *ex situ* conservation of crop diversity. However, post-CBD and subsequent initiatives (such as Gran Canaria Declaration – Anonymous, 2000; Global Strategy for Plant Conservation CBD – CBD, 2002a; European Plant Conservation Strategy – Anonymous, 2002; and the International Treaty on Plant Genetic Resources for Food and Agriculture which specifically focuses on agrobiodiversity – FAO, 2003) the shift, at least partially, to *in situ* conservation highlighted the lack of experience and appropriate techniques for its implementation that presented a methodological challenge to the conservation community.

The conservation of the full range of plant genetic diversity has historically often been associated with the conservation of socio-economically important species, because for these plant species the full range of genetic diversity is required for potential exploitation. These species are commonly regarded as a nation's plant genetic resources (PGR) that are equivalent in importance to a country's mineral or cultural heritage. PGR may be defined as *the genetic material of plants which is of value as a resource for the present and future generations of people* (IPGRI, 1993); and PGR for food and agriculture (PGRFA) are the PGR most directly associated with human food production and agriculture. PGRFA may be partitioned into six components: (i) modern cultivars; (ii) breeding lines and genetic stocks; (iii) obsolete cultivars; (iv) primitive forms of cultivated plants and landraces; (v) weedy races; and (vi) crop wild relatives (CWR). But it should be stressed that PGRFA is just one element or category of global or a country's plant genetic diversity (see Fig. 1.1). Modern cultivars, breeding lines, genetic stocks and obsolete cultivars are directly associated with modern breeding activities and constitute the bulk of gene bank holdings. Due to their location in breeding programmes or modern farming systems, their convenience of use by breeders and their rapid turnover *in situ* conservation is not applied to their conservation. But for socio-economically important species, *in situ* techniques are increasingly applied now to conserve landraces, weedy races and CWR species. The genetic diversity of these are generally regarded as being of less immediate breeding potential and therefore they are less well represented in gene banks. Landraces are traditional varieties of crops that have been maintained by farmers for millennia,

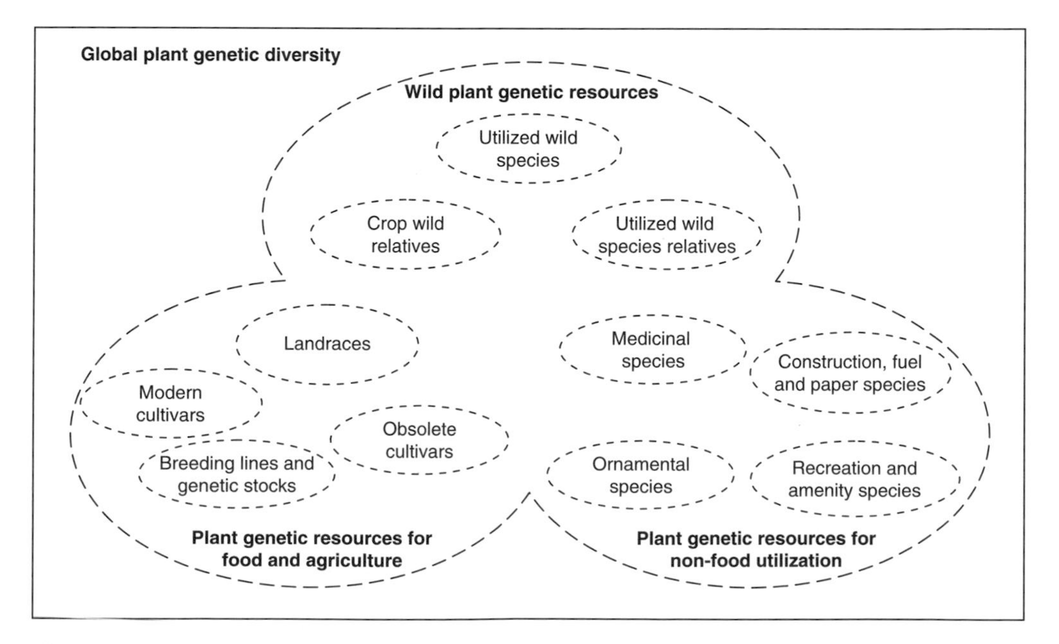

Fig. 1.1. Distinct categories of plant genetic diversity.

and as such they are not found in natural ecosystems. However, CWR are species that are more or less closely related to socio-economically important species and although having value associated with their potential as crop gene donors, are no different to any other wild species found in ecosystems worldwide.

Largely due to the sheer numbers of CWR species that exist, *ex situ* conservation has not been, and is not likely to be, a practical option, whereas *in situ* conservation offers the most pragmatic approach to conserving maximum CWR diversity for potential utilization. It is the conservation of the plant genetic diversity of these species that will be the prime focus of this volume, but it should be stressed that as CWR are in principle no different in terms of conservation to any other wild plant species, the techniques discussed in the following text will be equally applicable to wild plant species that are not regarded as CWR species.

CWR have been identified as a critical group vital for wealth creation, food security and environmental sustainability in the 21st century (Prescott-Allen and Prescott-Allen, 1983; Hoyt, 1988; Maxted *et al.*, 1997a; Meilleur and Hodgkin, 2004; Heywood and Dulloo, 2005; Stolton *et al.*, 2006). However, these species, like any other group of wild species, are subject to an increasing range of threats in their host habitats and appropriate protocols need to be applied to ensure humanity's exploitation options are maximized for future generations.

The refocusing of conservation activities onto *in situ* conservation along with the necessity of conserving the entire breadth of agrobiodiversity has challenged particularly the PGR community who had focused historically so extensively on *ex situ* techniques. Hawkes's (1991) comment that *in situ* PGR conservation techniques at the time of ratification of the CBD were in their infancy was very pertinent. Subsequently there has necessarily been a rapid progress in developing protocols and case studies for both the *in situ* conservation of crop landraces and CWR. With regard to the latter several useful texts have emerged, notably Horovitz and Feldman (1991), Gadgil *et al.* (1996), Maxted *et al.* (1997a), Tuxill and Nabhan (1998), Zencirci *et al.* (1998), Vaughan (2001), Heywood and Dulloo (2005), Stolton *et al.* (2006). In addition to these, the companion volume to this text *Crop Wild Relative Conservation and Use* (Maxted *et al.*, 2007) that arose from the EC-funded project 'Crop Wild Relative Diversity Assessment and Conservation Forum' (PGR Forum) was initiated specifically to address conservation issues related to CWR and broader *in situ* plant genetic diversity.

PGR Forum not only produced the first comprehensive CWR catalogue, the PGR Forum Crop Wild Relative Catalogue for Europe and the Mediterranean (Kell *et al.*, 2005; also see Kell *et al.*, 2007), but also investigated the production of baseline biodiversity data, the assessment of threat and conservation status for CWR, and generated methodologies for data management, population management and monitoring regimes, and for the identification and assessment of genetic erosion and genetic pollution; then it communicated these results to the broadest stakeholders, policy makers and user communities at the first International Conference on CWR held in Agrigento in September 2005. It should be stressed that although PGR Forum brought together European country partners with IUCN – The World Conservation Union and the International Plant Genetic Resources Institute (now Bioversity International) – the products are generic and can be applied in any country or region globally. As such, this publication is a product of PGR Forum and aims to provide practical protocols for the *in situ* conservation of CWR and other wild plant species, particularly focusing on the location, design, management and monitoring of plant genetic diversity within protected areas designated as genetic reserves.

1.2 What Are Crop Wild Relatives?

It is necessary to clarify what is meant by CWR as there is some debate within the scientific community. In the context of this publication, we regard CWR as those species relatively closely related to crops (or in fact any socio-economically valuable species), which may be crop progenitors and to which the CWR may contribute beneficial traits, such as pest or disease resistance, yield improvement or stability. They are generally defined in terms of any wild taxon belonging to the same genus as the crop (Plate 1). This definition is intuitively accurate and can be simply applied, but has resulted in the inclusion of a wide range of species that may not previously have been seen as particular CWR species. If the European and Mediterranean floras are taken as examples, approximately 80% of species can be considered CWR (Kell *et al.*, 2007). Therefore, there is a need to estimate the degree of CWR relatedness to enable limited conservation resources to be focused on priority species, those most closely related to the crop, easily utilized or severely threatened.

To establish the degree of crop relatedness one method would be to apply the Harlan and de Wet (1971) gene pool concept, close relatives being found in the primary gene pool (GP1) and more remote ones in the secondary gene pool (GP2). Interestingly, Harlan and de Wet (1971) themselves comment that GP2 may be seen as encompassing the whole genus of the crop and so may not restrict the number of CWR species included. This application of the gene pool concept remains functional for the crop complexes where hybridization experiments have been performed and the pattern of genetic diversity within the gene pool is well understood. However, for the majority of crop complexes, particularly in the tropics where species have been described and classified using a combination of morphological characteristics, the degree of reproductive isolation among species remains unknown and the application of the gene pool concept to define CWR is not possible. As a pragmatic solution, where there is a lack of crossing and genetic diversity data, the existing taxonomic hierarchy may be used (Maxted *et al.*, 2006). This can be applied to define a CWR's rank as follows:

- Taxon Group 1a – crop;
- Taxon Group 1b – same species as crop;
- Taxon Group 2 – same series or section as crop;
- Taxon Group 3 – same subgenus as crop;
- Taxon Group 4 – same genus;
- Taxon Group 5 – same tribe but different genus to crop.

Therefore, for CWR taxa where we have little or no information about reproductive isolation or compatibility, the Taxon Group concept can be used to establish the degree of CWR relatedness of a taxon. Although the application of the Taxon Group concept assumes that taxonomic distance is positively related to genetic distance, which need not be the case, on the whole the taxonomic hierarchy is likely to serve as a reasonable approximation of genetic distance and therefore, for practical purposes, classical taxonomy remains an extremely useful means of estimating genetic relationships. It is worth noting that while the Taxon Group concept can be applied to all crop and CWR taxa, the gene pool concept is understood for only approximately 22% of crop and CWR taxa (Maxted *et al.*, 2006).

As such, a CWR may be defined by pragmatic application of the gene pool and Taxon Group concepts to a crop and its wild relatives. A working definition of a CWR is thus provided by Maxted *et al.* (2006):

> A crop wild relative is a wild plant taxon that has an indirect use derived from its relatively close genetic relationship to a crop; this relationship is defined in terms of the CWR belonging to gene pools 1 or 2, or taxon groups 1 to 4 of the crop.

Therefore, taxa which belong to GP1B or TG1b and TG2 may be considered close CWR demanding higher priority for conservation, and those in GP2 or TG3 and TG4 more remote CWR affording lower priority. Those in GP3 and TG5 would be excluded from being considered CWR of that particular crop. Therefore, it can be argued that application of the gene pool and Taxon Group concepts to determine whether a species is or is not a CWR is pragmatic, and that the two concepts used together can be applied to establish the degree of CWR relatedness and thus assist in establishing conservation priorities.

Having both generally and more precisely defined a CWR, it needs to be restressed that the concept of a CWR is nominative, it is a human construct based on a wild species' potential use as a gene donor. As such, a CWR is intrinsically no different to any other wild plant species, and the fact that by extension from the Euro-Mediterranean region 80% of wild plant species are CWR means that most wild plants are CWR. This means, in terms of *in situ* conservation of plant genetic diversity, that the conservation of CWR and non-CWR species is synonymous and the techniques applied are equally applicable to both groups of plants.

1.3 Complementary PGR Conservation

It should be stressed that before wild plant taxa can be actively conserved *in situ* in a genetic reserve there are several steps that need to be taken. Maxted *et al.* (1997b) proposed an overall model for PGR conservation that sets genetic reserve within the context of the broader plant genetic conservation (see Fig. 1.2). As is shown, the decision must be taken as to whether the target taxon is of sufficient interest to warrant active conservation, an ecogeographic survey or a survey mission undertaken to identify appropriate hot spots of diversity, and specific conservation objectives generated and appropriate strategies outlined. The latter point must address the issue as to whether conservation in a genetic reserve is appropriate for the target taxon. If this is the case and the reserve is established successfully, a scheme that makes the conserved diversity available for current and future utilization must also be devised. The ultimate goal of genetic resources conservation is to ensure that the maximum possible genetic diversity of any taxon is maintained and available for potential utilization. PGR conservation is explicitly utilitarian in the sense that it acts as a link between the genetic diversity of a plant and its utilization or exploitation by humans as is shown in Fig. 1.2. Conservation and utilization are not two

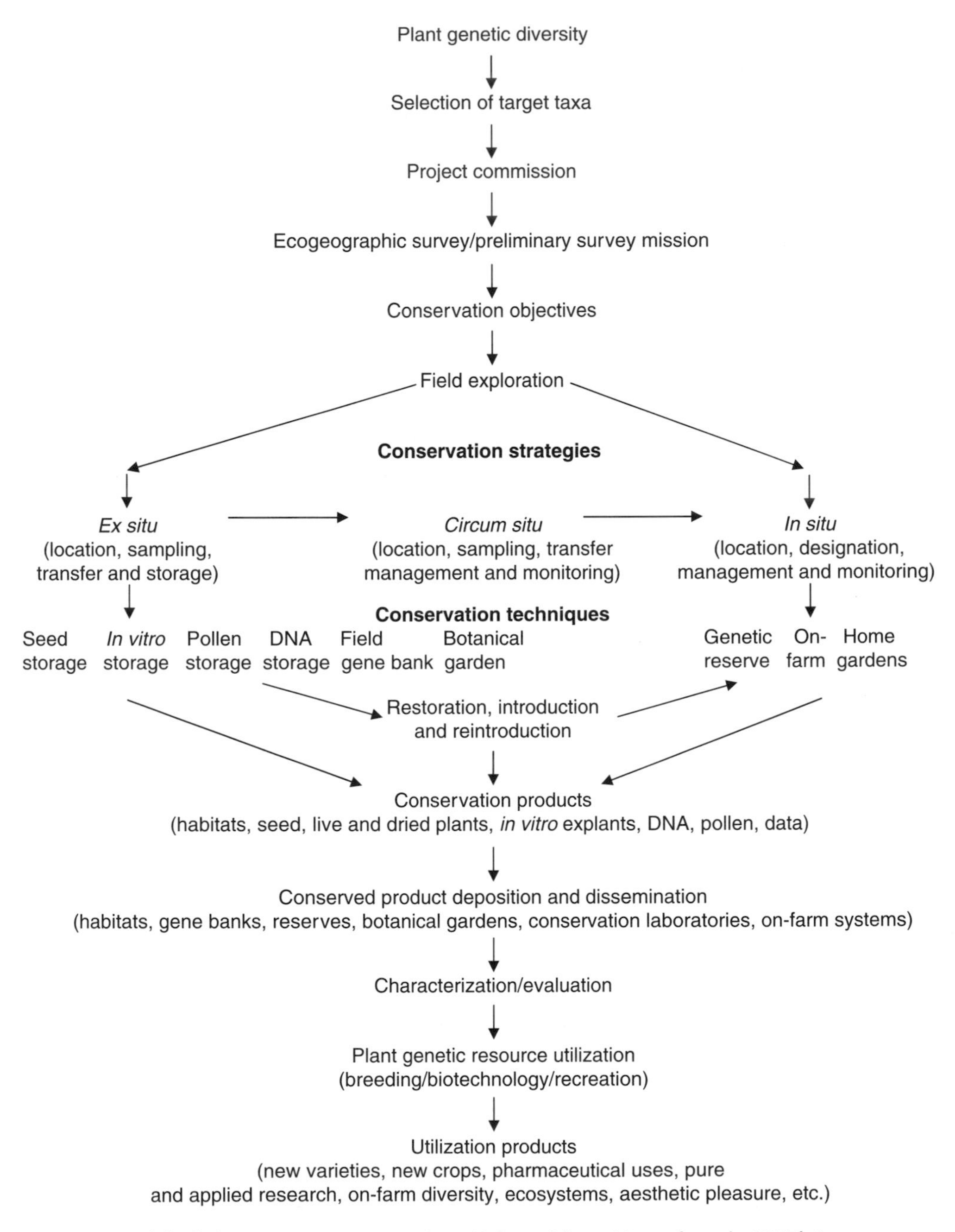

Fig. 1.2. Model of plant genetic conservation. (Adapted from Maxted *et al.,* 1997b.)

distinct end goals of working with plant diversity, but in fact are intimately linked (Maxted *et al.*, 1997b). Therefore, the model commences with the 'raw' material, plant genetic diversity, and concludes with the utilization products, and the component linking the steps is conservation.

As can be seen there are two fundamental strategies used in the conservation of PGR (Maxted *et al.*, 1997b):

- *Ex situ* – the conservation of components of biological diversity outside their natural habitats (CBD, 1992). The application of this strategy involves the location, sampling, transfer and storage of samples of the target taxa away from their native habitat (Maxted *et al.*, 1997b). Crop, CWR and wild plant species seeds can be stored in gene banks or in field gene banks as living collections. Examples of major *ex situ* collections include the International Maize and Wheat Improvement Center (CIMMYT) gene bank with more than 160,000 accessions (i.e. crop variety samples collected at a specific location and time); the International Rice Research Institute (IRRI), which holds the world's largest collection of rice genetic resources; and the Millennium Seed Bank at the Royal Botanic Gardens, Kew, which holds the largest collection of seed of 24,000 species primarily from global drylands.
- *In situ* – the conservation of ecosystems and natural habitats and the maintenance and recovery of viable populations of species in their natural surroundings and, in the case of domesticates or cultivated species, in the surroundings where they have developed their distinctive properties (CBD, 1992). *In situ* conservation involves the location, designation, management and monitoring of target taxa in the location where they are found (Maxted *et al.*, 1997b). There are relatively few examples of *in situ* genetic conservation for CWR species, but examples include *Zea perennis* in the Sierra de Manantlan, Mexico; *Aegilops* species in Ceylanpinar, Turkey; *Citrus*, *Oryza* and *Alocasia* species in Ngoc Hoi, Vietnam; and *Solanum* species in Pisac Cusco, Peru.

The goal of PGR conservation is to maximize the proportion of the gene pool of the target taxon conserved, whether *in situ* or *ex situ*, which can then be made available for potential or actual utilization. Both the application of *in situ* and *ex situ* techniques has its advantages and disadvantages as is shown in Table 1.1. However, the oft-cited major difference is that *ex situ* techniques freeze adaptive evolutionary development, especially that which is related to pest and disease resistance, while *in situ* techniques allow for natural genetic interactions between crops, their wild relatives and the local environment to take place. It should be acknowledged, however, that under extreme conditions of environmental change (such as local catastrophes or rapid climate change) extinction of genetic diversity rather than adaptation is likely to occur *in situ* (Stolton *et al.*, 2006). It is also fallacious to attempt cost comparisons between conservation strategies, as *in situ* conservation which is often cited as a 'cheap' option may be more costly if the target taxon requires more active management to maintain diversity. Management rarely focuses on single target taxon for *in situ* genetic conservation and it is likely that many wild plant species will be conserved in protected areas where they receive little or no direct conservation attention apart from monitoring provided the management regime has been accurately refined.

CBD Article 9 (CBD, 1992) stresses that the two conservation strategies (*ex situ* and *in situ*) cannot be viewed as alternatives or in opposition to one another but rather should be practised as complementary approaches to conservation. It is important where possible to apply a combination of both *in situ* and *ex situ* techniques so that they complement each other and conserve the maximum range of

Table 1.1. Summary of relative advantages and disadvantages of *in situ* and *ex situ* strategies. (Adapted from Maxted *et al.*, 1997.)

Strategy	Advantages	Disadvantages
Ex situ	1. Greater diversity of target taxa can be conserved as seed 2. Feasible for medium and long-term secure storage and disease resistance 3. Easy access for characterization and evaluation 4. Easy access for plant breeding and other forms of utilization 5. Little maintenance costs once material is conserved, except for field gene banks	1. Problems storing seeds of 'recalcitrant' species 2. Freezes evolutionary development, especially that which is related to pest 3. Genetic diversity may be lost with each regeneration cycle (but individual cycles can be extended to periods of 20–50 years or more) 4. *In vitro* storage may result in loss of diversity 5. Restricted to a single target taxon per accession (no conservation of associated species found in the same location)
In situ	1. Dynamic conservation in relation to environmental changes, pests and diseases 2. Provides easy access for evolutionary and genetic studies 3. Appropriate method for 'recalcitrant' species 4. Allows easy conservation of a diverse range of wild relatives 5. Possibility of multiple target taxa within a single reserve	1. Materials not easily available for utilization 2. Vulnerable to natural and man-directed disasters, e.g. climate change, fire, vandalism, urban development and air pollution 3. Appropriate management regimes remain poorly understood for some species 4. Requires high level of active supervision and monitoring 5. Limited genetic diversity can be conserved in any one reserve

genetic diversity (Maxted *et al.*, 1997b). Just because germplasm of a certain gene pool is maintained in a protected area and even though the site may be managed to maintain its diversity, it does not mean that the seed should not also be held in a gene bank or germplasm conserved using some other *ex situ* technique. Each complementary technique may be thought to slot together like pieces of a jigsaw puzzle to complete the overall conservation picture (Withers, 1993). The adoption of this holistic approach requires the conservationist to look at the characteristics and needs of the particular gene pool being conserved and then assess which of the strategies or combination of techniques offers the most appropriate option to maintain genetic diversity within that taxon.

1.4 *In Situ* PGR Conservation

The definition of *in situ* conservation used by the CBD (1992) instead of providing a general definition, as is the case for the definition of *ex situ* conservation, effectively conflates the definition of the two main *in situ* techniques that can be applied. A more generalized definition of *in situ* conservation would be *the conserva-*

tion of components of biological diversity in their natural habitats or traditional agroecological environments. This general definition of the *in situ* strategy may then be implemented using three types of techniques: protected area, on-farm and home garden conservation. It should be noted, as is discussed in Section 1.5, that protected area conservation is itself a broad term which encompasses several distinct applications and where the goal is to conserve genetic diversity within wild plant species and the *in situ* technique applied may be referred to as genetic reserve conservation.

Protected area and on-farm conservation are fundamentally distinct *in situ* applications, both in their targets (protected areas for wild species and on-farm for crops) and their management (protected areas are managed by conservationists and landraces conserved on-farm are managed by farmers). Home garden conservation may be seen as a variation of on-farm conservation, which is practised by non-commercial householders where the produce is consumed by the household.

- **Genetic reserves** (synonymous terms include genetic reserve management units, gene management zones, gene or genetic sanctuaries, crop reservations) – Involve the conservation of wild species in their native habitats. Genetic reserve conservation may be defined as *the location, management and monitoring of genetic diversity in natural wild populations within defined areas designated for active, long-term conservation* (Maxted *et al.*, 1997b). Practically this involves the location, designation, management and monitoring of genetic diversity within a particular, natural location. The site is actively managed even if that active management only involves regular monitoring of the target taxa. Also importantly the conservation is long-term, because significant resources will have been invested in the site to establish the genetic reserve and it would not be cost-effective to establish such a reserve in the short term. This technique is the most appropriate for the bulk of wild species, whether they are closely or distantly related to crop plants. If the management regime or management interventions are fairly minimal, it can be comparatively inexpensive, although still more expensive than *ex situ* gene bank conservation at US$5/year for a single accession (Smith and Linington, 1997). It is applicable for orthodox-seeded and non-orthodox-seeded species, permits multiple taxon conservation in a single reserve and allows for continued evolution. It is also important to make the point that genetic reserve conservation, as opposed to on-farm conservation and home garden conservation, is practised by professional conservationists, and so conservation is the prime concern (Plate 2).
- **On-farm conservation** – Involves the conserving of varieties within traditional farming systems and has been practised by traditional farmers for millennia. These farmers cultivate what are generally known as 'landraces'. Each season the farmers keep a proportion of harvested seed for re-sowing in the following year. Thus, the landrace is highly adapted to the local environment and is likely to contain locally adapted alleles or gene complexes. On-farm conservation may be defined as *the sustainable management of genetic diversity of locally developed landraces with associated wild and weedy species or forms by farmers within traditional agriculture, horticulture or agri-silviculture systems* (Maxted *et al.*, 1997b). The literature highlights a distinction in focus between at least two distinct, but associated, activities currently linked to on-farm conservation. The distinction between the

two is based on whether the focus is the conservation of genetic diversity within a particular farming system or the conservation of the traditional farming system itself, irrespective of what happens to the genetic diversity of landraces material within the farming system (Maxted *et al.*, 2002). These two variants of on-farm activities are obviously interrelated although may in certain cases be in conflict. For example, the introduction of a certain percentage of high-yielding varieties (HYVs) to a traditional farming system may sustain the farming system at that location, but could lead to gene replacement or displacement and therefore genetic erosion of the original landrace material. As such, where the focus is the conservation of genetic diversity within a particular farm it may be referred to as on-farm conservation, and where the focus is the conservation of the traditional farming system itself, as on-farm management.

- **Home garden management** – Crop on-farm conservation may be divided into field crop conservation where the crop is grown at least partly for external sale and more focused smaller scale home garden conservation where several crops are grown as small populations and the produce is used primarily for home consumption (Eyzaguirre and Linares, 2004). As such, home garden conservation may be regarded as a variation of on-farm conservation and may be defined as *the sustainable management of genetic diversity of locally developed traditional crop varieties by individuals in their backyard* (Maxted *et al.*, 1997b). Its focus is on medicinal, flavouring and vegetable species (e.g. tomatoes, peppers, coumarin, mint, thyme and parsley). Orchard gardens, which are often expanded versions of kitchen gardens, can be valuable reserves of genetic diversity of fruit and timber trees, shrubs, pseudo-shrubs such as banana and pawpaw, climbers and root and tuber crops as well as herbs.

1.5 Working Within Protected Areas

Protected areas, such as national parks, nature reserves and wilderness areas, may be broadly defined as areas set aside from development pressures to act as reservoirs for wild nature (Stolton *et al.*, 2006). Most protected areas were established to preserve exceptional geographical scenery or particular species or ecosystems, and are increasingly linked to global efforts at biodiversity conservation. However, there are very few known examples of protected areas established to specifically conserve CWR species (Hoyt, 1988; Maxted *et al.*, 1997a). In 2004, the Convention on Biological Diversity agreed upon a Programme of Work on Protected Areas, which aims to 'complete' ecologically representative protected area networks: systems of protected areas that contain all species and ecosystems in sufficient numbers and sufficiently large area to ensure their long-term survival. An additional justification for the completion of this initiative would be the preservation of socio-economically important CWR within these protected areas, which provides a strong augment for the enhancement of protected area networks.

It has been argued that CWR species are rarely associated with climax communities (Jain, 1975) and are therefore less likely to be found in protected areas which are commonly designated to conserve climax vegetation. However, this implies the application of a narrow definition of both CWR and protected areas (Stolton *et al.*, 2006). While the close CWR and progenitors of many of the major crops are more often associated with disturbed habitats, they are not exclusively so, and use of a broader definition of CWR will inevitably include species associated with the full range of habitats and successional stages. It is also mistaken to assume that protected areas are only established for climax communities; within all communities there are cyclical successional changes, and protected areas established near urban settlements are likely to be highly modified and have an intrinsic habitat disturbance dynamic. Therefore, protected areas contain a wealth of plants of direct or indirect socio-economic importance.

Forms of protected areas are very variable with diverse conservation goals and management regimes. IUCN (1995) defines a protected area as *an area of land and/or sea especially dedicated to the protection and maintenance of biological diversity, and of natural and associated cultural resources, and managed through legal or other effective means*, while the CBD (1992) defines a protected area as *a geographically defined area which is designated or regulated and managed to achieve specific conservation objectives*. IUCN (1995) identifies six distinct categories of protected areas depending on their management objectives (see Box 1.1).

As protected areas have not been established specifically to conserve the genetic diversity within CWR species, it is perhaps not surprising that none of the existing categories matches the definition of a genetic reserve outlined above. However, some of the existing IUCN categories are amenable for management adaptation to the conservation of the genetic diversity of wild plant species and CWR. Stolton *et al.* (2006) identify three categories as being most suitable:

- Category Ia – Strictly protected reserves (often small) set aside and left untouched to protect particular species under threat;
- Category II – Large ecosystem-scale protected areas maintained to allow CWR to continue to flourish and evolve under natural conditions;
- Category IV – Small reserves managed to maintain particular species, for example through controlled grazing or cutting to retain important grassland habitat, coppicing to maintain woodland ground flora or sometimes even intervening to restore habitat of threatened CWR species.

Although genetic reserves may be established in these protected areas, it would be preferable for an additional category to be added to the IUCN list that specifically addresses genetic reserve conservation.

Currently within protected areas the objective is likely to be broad biodiversity conservation at the ecosystem- or species-diversity level which may involve the detailed monitoring of keystone or indicator species, but is unlikely to focus on intraspecific diversity within any single species. As in the case of genetic conservation, the objective will be to maintain not only the appropriate effective population size, but also the level of genetic diversity within the target populations. As such, the management plan and regime for the site are likely to require adjustment to take this slightly different conservation focus into consideration. This might involve, in the case of weedy species, the maintenance of traditional agricultural practices or more active site management intervention to maintain the desired pre-climax vegetation.

Box 1.1. IUCN protected area management categories. (From IUCN, 1995.)

IUCN – The World Conservation Union has developed a definition and a series of categories of protected areas as outlined below.
Category Ia: ***Area managed mainly for science or wilderness protection*** – An area of land and/or sea possessing some outstanding or representative ecosystems, geological or physiological features and/or species, available primarily for scientific research and/or environmental monitoring.
Category Ib: ***Area managed mainly for wilderness protection*** – Large area of unmodified or slightly modified land and/or sea, retaining its natural characteristics and influence, without permanent or significant habitation, which is protected and managed to preserve its natural condition.
Category II: ***Area managed mainly for ecosystem protection and recreation*** – Natural area of land and/or sea designated to: (i) protect the ecological integrity of one or more ecosystems for present and future generations; (ii) exclude exploitation or occupation inimical to the purposes of designation of the area; and (iii) provide a foundation for spiritual, scientific, educational, recreational and visitor opportunities, all of which must be environmentally and culturally compatible.
Category III: ***Area managed mainly for conservation of specific natural features*** – Area containing specific natural or natural/cultural feature(s) of outstanding or unique value because of their inherent rarity, representativeness or aesthetic qualities or cultural significance.
Category IV: ***Area managed mainly for conservation through management intervention*** – Area of land and/or sea subject to active intervention for management purposes so as to ensure the maintenance of habitats to meet the requirements of specific species.
Category V: ***Area managed mainly for landscape/seascape conservation or recreation*** – Area of land, with coast or sea as appropriate, where the interaction of people and nature over time has produced an area of distinct character with significant aesthetic, ecological and/or cultural value, and often with high biological diversity. Safeguarding the integrity of this traditional interaction is vital to the area's protection, maintenance and evolution.
Category VI: ***Area managed mainly for the sustainable use of natural resources*** – Area containing predominantly unmodified natural systems, managed to ensure long-term protection and maintenance of biological diversity, while also providing a sustainable flow of natural products and services to meet community needs.

Just as protected areas encompass a range of different management types, so Stolton *et al.* (2006) conclude that they can have a number of different governance regimes and recognize four broad groupings of governance type (Borrini-Feyerabend *et al.*, 2004):

- **Government-managed protected areas** – Protected areas managed by national or local government, occasionally through an officially appointed independent body: i.e. federal or national ministry or agency in charge; local/municipal ministry or agency in charge; or government-delegated management (e.g. to an NGO);
- **Co-managed protected areas** – Protected areas which involve local communities in the management of government-designated protected areas through

active consultation, consensus-seeking, negotiating, sharing responsibility and transferring management responsibility to communities or NGOs, i.e. trans-boundary management, collaborative management (various forms of pluralist influence) or joint management (pluralist management board);

- **Private protected areas** – Protected areas managed by private individuals, companies or trusts, i.e. declared and run by an individual landowner, non-profit organization (e.g. NGO), university or cooperative or for-profit organization (e.g. individual or corporate landowners);
- **Community conserved areas** – Protected areas managed as natural and/or modified ecosystems voluntarily by indigenous, mobile and local communities.

The conservation of CWR species is appropriate under each regime as long as the site can be managed over a sustained period. It is also true that if any conservation project is to succeed in the long term, it must have the support of the local community; therefore, *in situ* conservation in community conservation areas does have the clear advantage of necessitating local support for the project. Experience has shown that once local communities realize they have a nationally, regionally or even globally valued resource in their local vicinity, they value it much more highly and this in itself will engender sustainability.

1.6 Genetic Reserve Conservation of Wild Plant Species

Within the context of *in situ* conservation of wild plant species genetic reserve conservation is the most appropriate conservation technique. Genetic reserves may be established on private lands, roadsides, in indigenous reserves and community-conserved areas as well as officially recognized protected areas; as such, it is important to note that they may equally well be established outside as inside protected areas (Plate 2). But often the simplest way forward in economic and political terms is for countries to locate genetic reserves in existing protected areas, e.g. national parks or heritage sites, as this is likely to provide some benefit to local people and so is likely to gain their support.

Practically, it could also be argued that *in situ* conservation of wild plant species in genetic reserves is the only practical option for their genetic conservation simply because of the need to conserve the full range of intraspecific genetic diversity and the sheer numbers of CWR that are involved. Kell *et al.* (2005, 2007) demonstrated that approximately 80% of the European and Mediterranean flora or 25,687 of the 30,983 plant species (Euro+Med PlantBase, 2005) are CWR species. Would there ever be sufficient resources available to conserve all these species and their intraspecific diversity *ex situ*? The answer seems unlikely to ever be positive and therefore the only realistic conservation option is *in situ* genetic reserve conservation, with *ex situ* conservation acting as an essential back-up system to ensure complementary conservation for the most important taxa. The Global Strategy for CWR Conservation and Use (Heywood *et al.*, 2007) recognizes this fact and recommends the identification at the regional, national and global level of a small number of priority sites (regional = 25, national = 5, global = 100) for the establishment of active CWR genetic reserves. These reserves would form an interrelated network of complementary internationally, regionally and nationally important CWR genetic reserve sites, and would also, if well selected, provide *in situ* conservation coverage for the broad genetic diversity of the species included.

However, it should be noted that many designated protected areas were established in less than ideal locations and/or are not actively managed, and as such do not offer adequate protection for biodiversity. The location of reserves is often practically dictated by the relative concentration of people and the suitability of the land for human exploitation (agriculture, urbanization, logging, etc.) and not because they are hot spots of biodiversity (Maxted *et al.*, 1997b). Primack (2006) cites two examples: the Greenland National Park, which is composed of a frozen land mass of 700,000 km^2, and the Bako National Park in Malaysia, which is set on nutrient-deficient soils, both of which are large in area but poor in biodiversity. In contrast, areas with high actual or potential economic value for human exploitation generally have fewer and smaller reserves. Given (1994) also illustrates the point further by listing the 15 largest reserves found in the USA, all of which are situated in agriculturally marginal areas. Although there may be a correlation between marginal lands and the lands that governments are willing to set aside to turn into protected areas, there is unlikely to be a natural correlation between marginal agricultural lands and the distribution of hot spots of biodiversity.

It can be argued that many existing protected areas are not actively managed for biodiversity conservation; in fact, it could even be argued that it is not possible to actively manage a site for all the biodiversity contained within it because of the competing management requirements of different species. Maxted *et al.* (1997b) distinguish between active and passive protected area conservation. Active management implies some form of dynamic intervention at the site, even if that intervention were simply limited to an agreement to monitor target populations included. Provided there is no deleterious change in the population levels, no further management intervention would be required. Whereas passive conservation involves less active intervention, by definition there is no management or monitoring of population, although there may be general ecosystem management, and all species are passively conserved if the entire ecosystem or habitat is stable, and individual species could be eroded and are inherently more vulnerable to extinction. It should therefore be understood that establishing genetic reserves in passively managed protected areas is likely to prove inefficient as the genetic conservation of plant diversity will certainly require active monitoring and management of the target plant populations.

It is also worth noting that many countries have now developed networks of less formal protected areas than those defined using the IUCN criteria (IUCN, 1995). These are often associated with agroenvironmental schemes, such as 'field margin programme' or 'conservation roadside verges' where often linear habitats that can be species-rich are specifically managed for conservation (Plate 5). In these cases, an incentive may be provided to the landowner, for example, to ensure that the site is not mown or grazed until after seeding of critical taxa or even a keystone species is planted to encourage the maintenance of diversity of a target species. However, the maintenance of these habitats and populations is under the control of the landowner and a change in owner or economic climate could result in management changes and negatively impact on the target species. Hence, agroenvironmental measures can produce short-term effects, but more formal nature conservation programmes, and the establishment of genetic reserves in existing protected areas, have a more stable long-term conservation basis.

Having made these points and accepting that no networks of protected areas offer ideal protection for all biodiversity, if the goal is *in situ* conservation of plant genetic diversity, experience thus far has shown that the establishment of genetic reserves is most efficient within existing actively managed protected areas. The reasons being: (i) these sites already have an associated long-term conservation ethos and are less prone to hasty management changes associated with private land or roadside where conservation value and sustainability are not a consideration; (ii) it is relatively easy to amend the existing site management to facilitate genetic conservation of wild plant species; and (iii) it means creating novel conservation sites can be avoided, thereby avoiding the possibly prohibitive cost of acquiring previously non-conservation-managed land.

Therefore, this volume will focus primarily on conserving plant genetic diversity within protected areas, where the species management is more directly under the control of the conservationists, but will also address *in situ* conservation of plant species outside of protected areas. Although it should be stressed that the definition of protected areas can include less formal conservation sites, such as roadsides, field margins or orchards, as well as more formal national parks, whether under state or private ownership, as long as the site is actively managed for conservation, it presents a potential site for the establishment of a genetic reserve.

As discussed, post CBD there was a need to develop practical *in situ* conservation methodologies. This need was recognized and first addressed for the genetic conservation of plants by Horovitz and Feldman (1991) and later by Maxted *et al.* (1997c) who proposed a methodology for *in situ* genetic reserve conservation (see Fig. 1.3). The model provides an overview to the procedure involved in planning, managing and using a genetic reserve, and it is upon these topics that subsequent chapters in this volume will build. The application of the model is also briefly summarized in Box 1.2. While it is recognized that this 'ideal' implementation is not always practically achievable, it is to what those establishing a genetic reserve might aspire.

1.7 *In Situ* Plant Genetic Diversity and Climate Change

It is argued throughout this text that the most appropriate approach to conservation of plant genetic diversity is the *in situ* genetic reserve approach, because of the sheer numbers of CWR taxa involved which prohibits general *ex situ* conservation, and it is felt desirable to maintain the co-evolutionary development of the CWR species within their biotic and abiotic environment, not to mention their valued contribution to general ecosystem maintenance. Underlying this proposition is the assumption that it is possible to conserve CWR genetic diversity for a long term *in situ*. However, there is a need to address the challenge of ecosystem change in the context of *in situ* CWR conservation.

It is now widely accepted that climate change is altering the geographic ranges of natural species and ecosystems (Walther *et al.*, 2002; Parmeson and Yohe, 2003).

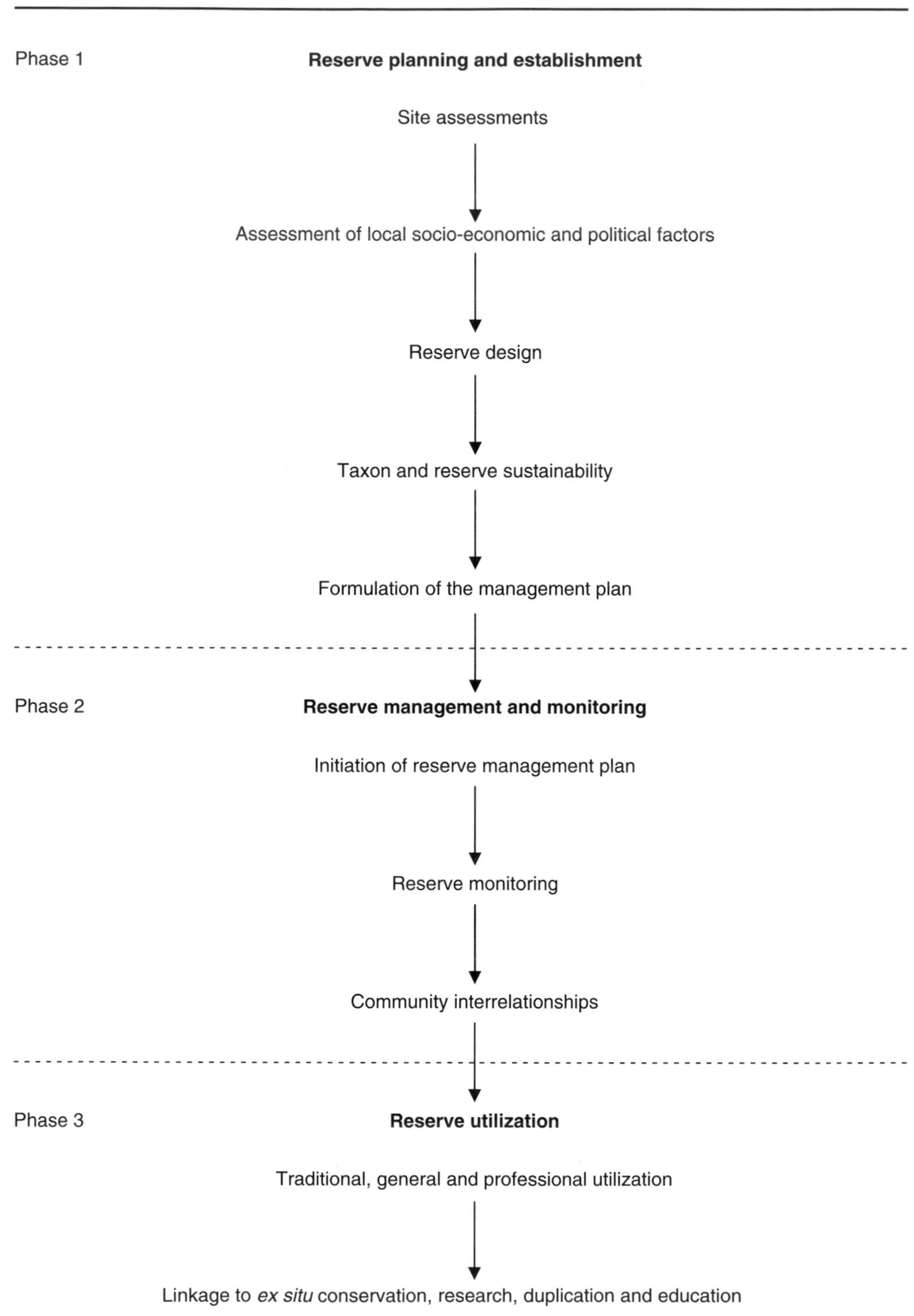

Fig. 1.3. Model for Genetic Reserve Conservation of CWR Species. (From Maxted *et al.*, 1997c.)

Box 1.2. Summary of procedure involved in *in situ* genetic conservation of wild plant species. (From Maxted *et al.*, 1997c.)

1. Selection of target taxa – Decide which species need active conservation and for which *in situ* genetic reserves is appropriate. If possible include more than one chosen species in each reserve.
2. Project commission – Formulate a clear, concise conservation statement establishing what species, why and in general terms where the species are to be conserved.
3. Ecogeographic survey/preliminary survey mission – This facilitates the collation of the basic information for the planning of effective conservation and surveys the distribution of taxonomic and genetic diversity, ecological requirements and the reproductive biology of the chosen species over its entire geographic range. Where little ecogeographic data are available, a preliminary course grid survey mission to collate the necessary background biological data on the species may be required.
4. Conservation objectives – Formulate a clear, concise set of conservation objectives, which state the practical steps that must be taken to conserve the species, and propose how the conserved diversity is linked to utilization.
5. Field exploration – Visit competing potential sites indicated as having high levels of target species and genetic diversity by the ecogeographic survey or preliminary course grid survey to 'ground truth' the predictions and identify specific locations where target species and genetic diversity are to be conserved in genetic reserves.
6. Conservation application for *in situ* genetic reserve – This involves the designation, management and monitoring of the genetic reserve.
6.1. Reserve planning and establishment
6.1.1. Site assessments – Within actual locations establish the sites where genetic reserves will be established; where possible they should cover the range of morphological and genetic diversity, and the ecological amplitude exhibited by the chosen species. Several reserves spread over the geographic range and the ecological environments occupied by the species may be required to cover a sufficiently large fraction of the target CWR species gene pool. Ensure that each reserve represents the fullest possible ecological range (micro-niches), to help secure maximal genetic variation, and to buffer the protected population against environmental fluctuations, pests and pathogens, and man-made disturbances. As part of this evaluation prepare a vegetation map of the area, surveying in detail the plant communities (and habitats) in which the target species grows.
6.1.2. Assessment of local socio-economic and political factors – Constraints ranging from economic to scientific and organizational will affect the establishment of the reserve. The simplest way forward in economic and political terms is for countries to take action on establishing a series of national parks or heritage sites, as this is likely to be of some benefit to the people of the countries and will gain their support.
6.1.3. Reserve design – Sites should be large enough to contain at least (1000–) 5000–10,000 individuals of each target species to prevent natural or anthropogenic catastrophes causing severe genetic drift or population unsustainability. Sites should be selected to maximize environmental heterogeneity. Each reserve site should be surrounded by a buffer zone of the same vegetation type, to facilitate immigration of individuals and gene flow, but also where experiments on management regimes might be conducted and visits by the public allowed, under supervision.
6.1.4. Taxon and reserve sustainability – Establishing and managing an *in situ* genetic reserve is resource-expensive and therefore both the taxon and reserve must be deemed sustainable over an extended period of time or the investment will be forfeited.

Continued

Box 1.2. *Continued*

6.1.5. Formulation of the management plan – The reserve site would have been selected because it contained abundant and hopefully genetically diverse populations of the target taxon. Therefore, the first step in formulating the management plan is to observe the biotic and abiotic qualities and interactions at the site. Once these ecological dynamics within the reserve are known and understood, a management plan that incorporates these points, at least as they relate to the target taxon, can be proposed.

6.2. Reserve management and monitoring

6.2.1. Initiation of reserve management plan – It is unlikely that any management plan will be wholly appropriate when first applied; it will require detailed monitoring of target and associated taxa and experimentation with the site management before a more stable plan can evolve. The plan may involve experimentation with several management interventions (a range of grazing practices, tree-felling, burning, etc.) within the reserve to ensure that the final plan does meet the conservation objectives, particularly in terms of maintaining the maximum CWR species and genetic diversity. Genetic reserves conservation is a process-oriented way of maintaining genetic resources; it will maintain not only the evolutionary potential of a population but also the effective population sizes of the CWR species.

6.2.2. Reserve monitoring – Each site should be monitored systematically at a set time interval and the results fed back in an iterative manner to enhance the evolving management regime. The monitoring is likely to take the form of measures of CWR taxon number, diversity and density as measured in permanent transects, quadrats, etc.

6.3. Reserve utilization

6.3.1. Traditional, general and professional utilization – Humans generally conserve because they wish to have actual or potential utilization options; therefore, when designing the reserve it is necessary to make an explicit link between the material conserved and that currently or potentially utilized by humankind. There are three basic user communities: traditional or local, the general public and professional users.

6.3.2. Linkage to ex situ *conservation, duplication, research and education* – There is a need to form links with *ex situ* conserved material to ensure utilization and also as a form of back-up safety duplication. The reserve forms a natural platform for ecological and genetic research, as well as providing educational opportunities for the school, higher educational and general public levels.

7. Conservation products – These will be populations of live plants held in the reserve, voucher specimens and the passport data associated with the reserve and plant populations.

8. Conserved product deposition and dissemination – The main conserved products, the plant populations of the target taxon, are held in the reserve. However, there is a need for safety duplication and a sample of germplasm should also be periodically sampled and deposited in an appropriate *ex situ* collection (gene bank, field gene bank, *in vitro* banks, botanical gardens or conservation laboratory) with the appropriate passport data.

9. Characterization/Evaluation – The first stage of utilization will involve the recording of genetically controlled characteristics (characterization) and the material may be grown out under diverse environmental conditions to evaluate and screen for

Continued

Plate 1

Plate 2

Plate 1 *Lupinus gredensis*
Lupinus is a genus of the legume family that has about 200 species which originated in the Mediterranean Region (subgenus *Lupinus*) and America (subgenus *Platycarpos*). Most *Lupinus* species have seeds with a high protein content used for cattle and also for human consumption (*L. albus* and *L. mutabilis*). The largest producer of cultivated *Lupinus* is Australia (over 1 million t), which is far away from the centres of speciation. The photograph shows plants of *Lupinus gredensis*, a crop wild relative, growing on an abandoned field 20 km NE of Madrid, Spain. (Photo credit: Lori De Hond)

Plate 2 Wild and cultivated fruit trees
The Wadi Sair Genetic Reserve was established near Hebron in 1995 as part of the GEF funded "Conservation and sustained use of dryland agro-biodiversity" project. The valley reserve contains fruit tree, forage legume and some vegetable crop wild relatives, as well as cultivated fruit trees. (Photo credit: Nigel Maxted)

Plate 3

Plate 4

Plate 3 Plant microreserve
Decisions on reserve boundaries must be made with detailed information of the target species (census, population trends) and the habitat. A plant officer gathers phytosociological data on the local vegetation to design the buffer area for a plant microreserve in Llombai (Valencian Community, Spain), to conserve populations of the recently discovered *Lupinus mariae-josephi*. (Photo credit: Emilio Laguna)

Plate 4 Link with local communities
An important component of successful crop wild relative conservation is establishing the link between local communities and plant genetic diversity. Here a local farmer is being questioned about the management of field margins rich in crop wild relatives near Kenitra (Morocco). (Photo credit: Nigel Maxted)

Box 1.2. *Continued*

drought or other tolerance, or the experimental infection of the material with diseases or pests to screen for particular biotic resistance (evaluation).
10. Plant genetic resource utilization – The conserved material is likely to be used in breeding and biotechnology programmes, provide food, fuel, medicines, industrial products, as well as a source of recreation and education. Locally the materials held in the reserve may have traditionally been used in construction, craft, adornment, transport or food. This form of traditional utilization of the reserve by local people should be encouraged, provided it is sustainable and not deleterious to the target taxon or taxa, as it is essential to have local support for conservation actions if the reserve is to be sustainable in the medium to long term.

It is also affecting the phenological cycles of species. Studies have shown that the majority of species show trends towards earlier flowering and budburst (e.g. Parmeson and Yohe, 2003). Also, increases in productivity in some species as a result of elevated atmospheric CO_2 will be negated by the impacts of higher temperatures (Batts *et al.*, 1998), which lead to a shortened life cycle, accelerated development and reduced seed production and fertility (Wollenweber *et al.*, 2003).

The *in situ* conservation of wild plant species will be negatively affected because, particularly for those species with restricted distributions or narrow climatic envelopes, they may not have the ability to migrate or adapt to changing climatic conditions. The species either need to have appropriate genetic diversity to adapt to the novel condition or need to migrate with their climatic envelope. However, it is uncertain to what extent either of these conditions may be met by individual species, although the characteristics likely to be associated with susceptibility may be predicted, as will be discussed in later chapters. It is likely that those species without sufficient adaptive amplitude or with a limited capacity to migrate to appropriate homoclines are likely to go extinct, possibly even within the reserves established to protect them. Habitat fragmentation resulting from the spatially heterogeneous effects of climate change also will impact on the genetic viability of populations, further increasing species' vulnerability to genetic erosion. Species with currently narrow distributions are predicted to be especially vulnerable to climate change (Schwartz *et al.*, 2006), and it is likely that climate change will force even more species into narrower ranges.

For these reasons, the expected impacts of climate change and species' responses must be considered at all stages of *in situ* plant conservation, from choice of species for inclusion in a genetic reserve to the management and monitoring of the genetic reserves itself. To ignore them will undermine the considerable investments required for effective conservation and, more importantly, risk degradation and loss of genetic diversity. Therefore, this important contemporary issue will be addressed throughout this text.

1.8 Conclusions

There has been a growing interest among genetic conservationists in the *in situ* plant conservation, and various projects such as the EC Framework 5-funded project,

PGR Forum and several GEF-funded projects have made significant advances in the creation of national and regional CWR Catalogue, CWR information management and conservation techniques for CWR species. Apart from these practical scientific achievements, equally important has been the raising of the profile of CWR conservation and use within both the public and professional communities. This volume aims to provide practical protocols for the *in situ* conservation of wild plants that are globally applicable. The methodologies outlined are derived in part from those established by ecologically based conservationists over the last century but are specifically adapted for application in the conservation of plant genetic diversity context. There remain few detailed worked examples of *in situ* conservation for CWR species and few genetic reserves have been established for PGR conservation; however, components of the protocols being proposed have been tested and it is hoped that they will prove useful for those charged with *in situ* conservation for CWR. The achievements have only been possible due to the collaborative efforts of a network of committed individuals (both PGR Forum partners and collaborators alike) who have the common aim of conserving these vital resources.

But in many ways, both within Europe and globally, the work is only just beginning. There is now a need to enact the recommendations that follow, to ensure genetic reserves are established and, given the resources and legal protection, to ensure they are sustainable. There is a need to complete and implement the policy targets outlined in the Global Strategy for CWR Conservation and Use and to strengthen global collaborative efforts via the newly established IUCN/SSC CWR Specialist Group, through the ECP/GR *in situ* and on-farm network and the work of other CWR projects globally.

To conclude, the growing interest throughout the world (particularly in the light of recent biotechnological advancement) in the wealth of CWR diversity for exploitation will only be justified if the conserved diversity is made available to the user community. Although until recently CWR species have been sporadically conserved *ex situ* and rarely actively conserved *in situ*, PGR Forum and other initiatives have in recent years made significant progress in raising consciousness of the need to conserve CWR and in developing the foundations and protocols necessary for efficient and effective conservation both in Europe and globally. The need for such protocols globally is particularly prescient in the context of continued threats to genetic diversity from genetic erosion and extinction, not least in the face of rapid ecosystem change led by the impact of climate change. This has been recognized by the Conference of the Parties (COP) to the CBD 2010 Biodiversity Target (CBD, 2002b) where parties are committed ‘to achieve by 2010 a significant reduction of the current rate of biodiversity loss at the global, regional and national level as a contribution to poverty alleviation and to the benefit of all life on earth’. For us to be able to address this target, along with the requirements of other relevant international, regional and national strategies and legislation, we need to have firm knowledge of what natural plant genetic diversity exists, be able to assess changes over time and specifically ensure that we can effectively and efficiently conserve this diversity so that it is available for possible exploitation by future generation.

References

Anonymous (2000) *Gran Canaria Declaration: Calling for a Global Programme of Plant Conservation*. Botanic Gardens Conservation International, Kew, Surrey, UK.

Anonymous (2002) *European Plant Conservation Strategy*. Council of Europe and Planta Europa, London.

Batts, G.R., Ellis, R.H., Morison, J.I.L., Nkema, P.N., Gregory, P.J. and Hadley, P. (1998) Yield and partitioning in crops of contrasting cultivars of winter wheat in response to CO_2 and temperature in field studies using temperature gradient tunnels. *Journal of Agricultural Sciences, Cambridge* 130, 17–27.

Borrini-Feyerabend, G., Kothari, A. and Oviedo, G. (2004) *Indigenous and Local Communities and Protected Areas: Towards Equity and Enhanced Conservation*. IUCN, Gland,Switzerland/Cambridge.

Convention on Biological Diversity (1992) *Convention on Biological Diversity: Text and Annexes*. Secretariat of the Convention on Biological Diversity, Montreal, Canada.

Convention on Biological Diversity (2002a) *Global Strategy for Plant Conservation*. Secretariat of the Convention on Biological Diversity, Montreal, Canada. Available at: http://www.biodiv.org/decisions/?lg=0&dec=VI/9.

Convention on Biological Diversity (2002b) *2010 Biodiversity Target*. Secretariat of the Convention on Biological Diversity, Montreal, Canada. Available at: http://www.biodiv.org/2010-target/default.aspx.

Euro+Med PlantBase (2005) *Euro+Med PlantBase: The Information Resource for Euro-Mediterranean Plant Diversity*. Dipartimento di Scienze botaniche ed Orto botanico, Università degli Studi di Palermo. Available at: http://www.emplantbase.org/home.html.

Eyzaguirre, P.B. and Linares, O.F. (eds) (2004) *Home Gardens and Agrobiodiversity*. Smithsonian Books, Washington, DC.

Food and Agriculture Organisation (2003) *International Treaty on Plant Genetic Resources for Food and Agriculture*. Food and Agriculture Organization of the United Nations, Rome, Italy.

Gadgil, M., Niwas Singh, S., Nagendra, H. and Chandran, M.D.S. (1996) In Situ *Conservation of Wild Relatives of Cultivated Plants: Guiding Principles and a Case Study*. Food and Agriculture Organization of the United Nations, Rome, Italy.

Given, D.R. (1994) *Principles and Practice of Plant Conservation*. Chapman & Hall, London.

Harlan, J.R. and de Wet, J.M.J. (1971) Towards a rational classification of cultivated plants. *Taxon* 20, 509–517.

Hawkes, J.G. (1991) International workshop on dynamic *in situ* conservation of wild relatives of major cultivated plants: summary of final discussion and recommendations. *Israel Journal of Botany* 40, 529–536.

Heywood, V.H. and Dulloo, M.E. (2005) In Situ *Conservation of Wild Plant Species: A Critical Global Review of Good Practices*. IPGRI Technical Bulletin No. 11. International Plant Genetic Resources Institute, Rome, Italy.

Heywood, V.H., Kell, S.P. and Maxted, N. (2007) Towards a global strategy for the conservation and use of crop wild relatives. In: Maxted, N., Ford-Lloyd, B.V., Kell, S.P., Iriondo, J.M., Dulloo, E. and Turok, J. (eds) *Crop Wild Relative Conservation and Use*. CAB International, Wallingford, UK.

Horovitz, A. and Feldman, M. (1991) Population dynamics of the wheat progenitor, *Triticum turgidum* var. *dioccoides*, in a natural habitat in Eastern Galilee. *Israel Journal of Botany* 40(5–6), 349–536.

Hoyt, E. (1988) *Conserving the Wild Relatives of Crops*. IBPGR, IUCN, WWF, Rome/Gland, Switzerland.

IPGRI (1993) *Diversity for Development*. International Plant Genetic Resources Institute, Rome, Italy.

IUCN Commission on National Parks and Protected Areas with the World Conservation Monitoring Centre (1995) *Guidelines for Protected Area Management Categories*. IUCN, Gland, Switzerland.

Jain, S.K. (1975) Genetic reserves. In: Frankel, O.H. and Hawkes, J.G. (eds) *Crop Genetic Resources for Today and Tomorrow*.

Cambridge University Press, Cambridge, pp. 379–396.

Kell, S.P., Knüpffer, H., Jury, S.L., Maxted, N. and Ford-Lloyd, B.V. (2005) *Catalogue of Crop Wild Relatives for Europe and the Mediterranean*. Available at: http://cwris.ecpgr.org/ and on CD-ROM. University of Birmingham, Birmingham, UK.

Kell, S.P., Knüpffer, H., Jury, S.L., Ford-Lloyd, B.V. and Maxted, N. (2007) Crops and wild relatives of the Euro-Mediterranean region: making and using a conservation catalogue. In: Maxted, N., Ford-Lloyd, B.V., Kell, S.P., Iriondo, J.M., Dulloo, E. and Turok, J. (eds) *Crop Wild Relative Conservation and Use*. CAB International, Wallingford, UK.

Maxted, N., Ford-Lloyd, B.V. and Hawkes, J.G. (eds) (1997a) *Plant Genetic Conservation: The* In Situ *Approach*. Chapman & Hall, London.

Maxted, N., Ford-Lloyd, B.V. and Hawkes, J.G. (1997b) Complementary conservation strategies. In: Maxted, N., Ford-Lloyd, B.V. and Hawkes, J.G. (eds) *Plant Genetic Conservation: The* In Situ *Approach*. Chapman & Hall, London, pp. 20–55.

Maxted, N., Hawkes, J.G., Ford-Lloyd, B.V. and Williams, J.T. (1997c) A practical model for *in situ* genetic conservation. In: Maxted, N., Ford-Lloyd, B.V. and Hawkes, J.G. (eds) *Plant Genetic Conservation: The* In Situ *Approach*. Chapman & Hall, London, pp. 545–592.

Maxted, N., Guarino, L., Myer, L. and Chiwona, E.A. (2002) Towards a methodology for on-farm conservation of plant genetic resources. *Genetic Resources and Crop Evolution* 49, 31–46.

Maxted, N., Ford-Lloyd, B.V., Jury, S.L., Kell, S.P. and Scholten, M.A. (2006) Towards a definition of a crop wild relative. *Biodiversity and Conservation* 15(8), 2673–2685.

Maxted, N., Kell, S.P. and Ford-Lloyd, B. (2007) Crop wild relative conservation and use: establishing the context. In: Maxted, N., Dulloo, E., Ford-Lloyd, B., Kell, S.P. and Iriondo, J.M. (eds) *Crop Wild Relatives Conservation and Use*. CAB International, Wallingford, UK.

Meilleur, B.A. and Hodgkin, T. (2004) *In situ* conservation of crop wild relatives: status and trends. *Biodiversity and Conservation* 13, 663–684.

Parmesan, C. and Yohe, G. (2003) A globally coherent fingerprint of climate change impacts across natural systems. *Nature* 421, 37–42.

Prescott-Allen, R. and Prescott-Allen, C. (1983) *Genes from the Wild: Using Wild Genetic Resources for Food and Raw Materials*. Earthscan, London.

Primack, R.B. (2006) *Essentials of Conservation Biology*, 4th edn. Sinauer Associates, Sunderland, Massachusetts.

Schwartz, M.W., Iverson, L.R., Prasad, A.M., Matthews, S.N. and O'Connor, R.J. (2006) Predicting extinctions as a result of climate change. *Ecology* 87(7), 1611–1615.

Smith, R.D. and Linington, S. (1997) The management of the Kew Seed Bank for the conservation of arid land and UK wild species. *Bocconea* 7, 273–280.

Stolton, S., Maxted, N., Ford-Lloyd, B., Kell, S.P. and Dudley, N. (2006) *Food Stores: Using Protected Areas to Secure Crop Genetic Diversity*. WWF Arguments for Protection series. Gland, Switzerland.

Tuxill, J. and Nabhan, G. (1998) *Plants and Protected Areas: A Guide to* In Situ *Management*. People and Plants Conservation Manual 3. Stanley Thornes, Cheltenham, UK.

Vaughan, D. (2001) In Situ *Conservation Research*. MAFF, Tsukuba, Japan.

Walther, G.-R., Post, E., Convey, P., Menzel, A., Parmesank, C., Beebee, T.J.C., Fromentin J.-M., Hoegh-Guldberg, O. and Bairlein, F. (2002) Ecological responses to recent climate change. *Nature* 416, 389–395.

Withers, L.A. (1993) Conservation methodologies with particular reference to *in vitro* conservation. In: Ramanatha Rao, V. (ed) *Proceedings of the Asian Sweet Potato Germplasm Network Meeting*. Guangzhou, China. CIP, Manila, The Philippines, pp. 102–109.

Wollenweber, B., Porter, J.R. and Schellberg, J. (2003) Lack of interaction between extreme high-temperature events at vegetative and reproductive growth stages in wheat. *Journal of Agronomy and Crop Science* 189, 142–150.

Zencirci, N., Kaya, Z., Anikster, Y. and Adams, W.T. (1998) *Proceedings of International Symposium on* In Situ *Conservation of Plant Diversity*. Central Research Institute for Field Crops, Ankara, Turkey.

2 Genetic Reserve Location and Design

M.E. Dulloo,[1] J. Labokas,[2] J.M. Iriondo,[3] N. Maxted,[4] A. Lane,[1] E. Laguna,[5] A. Jarvis[6] and S.P. Kell[4]

[1]*Bioversity International, Rome, Italy;* [2]*Institute of Botany, Vilnius, Lithuania;* [3]*Área de Biodiversidad y Conservación, Depto. Biología y Geología, ESCET, Universidad Rey Juan Carlos, Madrid, Spain;* [4]*School of Biosciences, University of Birmingham, Edgbaston, Birmingham, UK;* [5]*Centro para la Investigación y Experimentación Forestal (CIEF), Generalitat Valenciana, Avda. País Valencià, Valencia, Spain;* [6]*Bioversity International and International Centre for Tropical Agriculture c/o CIAT, Cali, Colombia*

2.1 Introduction

Effective *in situ* conservation of crop wild relatives (CWR) requires that maximum genetic diversity of the targeted species be adequately represented and sustainably managed in protected areas. In most cases, one single protected area would not be sufficient to fully conserve the desired extent of diversity, except when dealing with very narrow endemics; therefore, several sites would be needed. It is thus important to work towards a network of genetic reserves for selected CWR, already located as far as possible in protected areas. However, there will be instances where populations located outside protected areas would have to be considered. If *in situ* protection cannot be afforded for valid reasons or cases where there is evidence of a quick decline in the abundance of the target species in well-protected *in situ* areas (e.g. the case of *Ulmus minor* around Europe), complementary *ex situ* conservation actions would be needed to conserve the populations (see Engels *et al.*, Chapter 6, this volume). There are two key factors that influence the effective conservation of target species: the proper selection of the best sites for genetic reserves (i.e. the location of reserves) and the design of such reserves (reserve design).

In this chapter, procedures for selecting genetic reserves for target species and their design are discussed and guidelines provided. It is important to realize that by their very nature 'guidelines' are simple rough indications on how to proceed in a particular task. There can be no single specific procedure for selecting and designing genetic reserves that will fit all cases due to the great diversity of situations that may arise. On the other hand, reserve selection and systematic planning usually apply to a multi-species approach where the objectives are the conservation of maximum species or ecosystem diversity. However, in this chapter we take a taxon-specific approach. We are more concerned about selection and design of genetic reserve areas for single taxa of CWR and other wild species.

Among the main points in establishing genetic reserves are their location, size and number. The procedures need to reflect what is known about the individual species' geographical, ecological and physiological (including reproductive biology) attributes. In identifying and designing conservation areas, management policies or actions should always recognize that each and every individual of a species in nature is genetically unique. Each species has its own distribution or occurrence pattern and may exhibit distinct phases during its life cycle. Species may also have unique symbiotic or commensal associations with other biotic components of the ecosystem, e.g. pollinators, hosts and dispersers. Criteria for prioritizing species and identifying sites for reserve establishment need to consider these attributes as well as activities of both human and other biotic components taking place within, and in the vicinity of, the reserves. These have a major influence on the way reserves are designed. The choice should be made of what is more effective – single large or several small reserves. For this purpose several groups of factors should be considered: natural (biological, ecogeographic, climatic); socio-economic (human activities, use of target taxa and/or land); legal (legal status of species and types of natural protected sites); and political (national laws, land-use policies, international protocols and treaties). International, regional and national policies, legislation and conventions governing protected areas and biodiversity need to ensure the long-term monitoring and management of genetic reserves. These aspects dictate the feasibility and sustainability of putative genetic reserves, and can greatly influence the final decisions on their locations and design. Drawing parallels with other categories of protected areas, an account should be made on the peculiarities of this particular category of protected areas, where conservation of genetic material is being achieved through active management of a reserve and sustainable germplasm utilization.

2.2 Genetic Reserve Location

The objective of genetic reserve location is to determine which areas or sites containing the target species are the most important in terms of genetic diversity for the creation of a genetic reserve or a network of genetic reserves. Since the focus is on genetic diversity within a single taxon, the basic management unit for conservation must be below the species level. Diversity could be considered at the population, genetic or allelic level. For all practical purposes in the selection of reserves for targeted species, the population should be the management unit, given that populations are most widely monitored, exploited and managed by people (see Iriondo *et al.*, Chapter 4, this volume). In fact the conservation of populations of one or a few species used to be the main reason to declare new protected areas (Simberloff, 1986). The status of populations is also often used as a proxy to provide insight into the status of genetic diversity, as the extinction of unique populations may represent the loss of unique genetic diversity contained in those species and populations (Millennium Ecosystem Assessment, 2005). The methodology used for identifying genetic reserves influences the scope and the cost of the exercise and will determine whether the range of the genetic reserve is representative of the genetic diversity of the target species.

The most important prerequisite for the proper selection of genetic reserve locations is adequate knowledge of the target species and their habitats. This knowledge will allow conservation planners to select the most optimum sites for inclusion in a genetic reserve network system. The information required may be broadly divided into five categories:

1. Taxonomic;
2. Demographic;
3. Genetic;
4. Ecological;
5. Policy and socio-economics.

In addition to the above, information on physiology (including phenotypic plasticity), morphology and reproductive biology should also be taken into consideration, but these may be more important when designing reserves (see Section 2.5).

One of the most commonly used techniques for gathering information for generating this knowledge base is ecogeographic survey. This technique provides a useful way of collating taxonomic, genetic, ecological and geographical information about a target taxon to help identify key areas for *in situ* conservation. The concept was first elaborated in 1985 by the International Board for Plant Genetic Resources (IBPGR) (renamed as International Plant Genetic Resources Institute (IPGRI) in 1992 and Bioversity International in 2006) with the view to improve the *in situ* conservation of CWR (IBPGR, 1985). The report of the IBPGR taskforce (IBPGR, 1985) showed the value of gathering ecogeographic information in locating significant genetic material. Thus, representative populations can be monitored to guide the selection of representative samples for conservation and utilization (IBPGR, 1985). Since then, the concept has been further refined and Maxted *et al.* (1995) provided procedures for undertaking an ecogeographical survey or study, recognizing that a full study is not always feasible and requires significant time and financial resources. However, much information can still be collated with limited resources to enable informed decisions to be made about target taxa.

Maxted *et al.* (1995) provide a detailed description of the different steps for carrying out an ecogeographic survey or study. This is illustrated in Fig. 2.1. It basically consists of three phases, namely: project design, data collection and analysis, and product generation. Project design involves the collation and analysis of large and complex data sets obtained from literature, herbaria, gene banks and other institutions which hold germplasm, as well as people (both specialists and local people) who have much scientific and traditional knowledge on the target species and their populations as mentioned above. The ecogeographic data analysis produces three basic products: (i) a database which contains the raw data for each taxon; (ii) the conspectus which summarizes the data for each taxon; and (iii) the report which discusses the contents of the database and conspectus (Maxted *et al.*, 1995; Maxted and Kell, 1998).

2.2.1 Taxonomic information

First of all, the identity of the target taxon needs to be clearly established, as this will ultimately have consequences for the selection of reserve areas (STAP-GEF, 1999). The incorrect identification of a taxon may lead to the selection of specific sites where the target taxon does not actually occur. Intraspecific variation is of major importance in selecting sites for genetic reserves since it is often the basis for distinguishing subspecies, ecotypes or chemotypes. Tolerance traits and resistance to a particular disease are often traced to a small number of plants in a very specific region, such as resistance to barley yellow dwarf virus originating in the highlands of Ethiopia, and resistance to Russian wheat aphid from Iran and neighbouring countries (Nevo, 1988). Proper taxonomic information as well as common names used by local communities should, therefore, be well documented. Standard Floras provide a reliable source of taxonomic information on plants which occur in a given country or region and should be used for their identification and the nomenclature adopted therein should be followed, unless it is possible to determine the correct name (if different) through other sources (Heywood and Dulloo, 2005). For example, lists of Standard Floras exist for Europe (Tutin *et al.*, 1964–1988, 1993) and the Mediterranean region (Heywood, 2003) as well as Euro+Med PlantBase (2002) for the combined Euro-Mediterranean region. In addition, regional treatments such as the *Flora Europaea* (Tutin *et al.*, 1964–1988, 1993) and *Med-Checklist* (Greuter *et al.*, 1989) are available. A comprehensive taxonomic database and information system for the combined region is in an advanced state of preparation (see http://www.euromed.org.uk). There are standard lists of the floristic works for each country and region (Davis *et al.*, 1986; Frodin, 2001) which show that wherever an ecogeographic survey or study is undertaken there are detailed floristic data available. However, in practice when working in less well-studied regions of the world than Europe, these floras may be historic and difficult to access easily. Additional ecogeographic data may be derived from specific revisions or monographic treatments of the target taxon, and these may provide particularly detailed sources of intraspecific diversity.

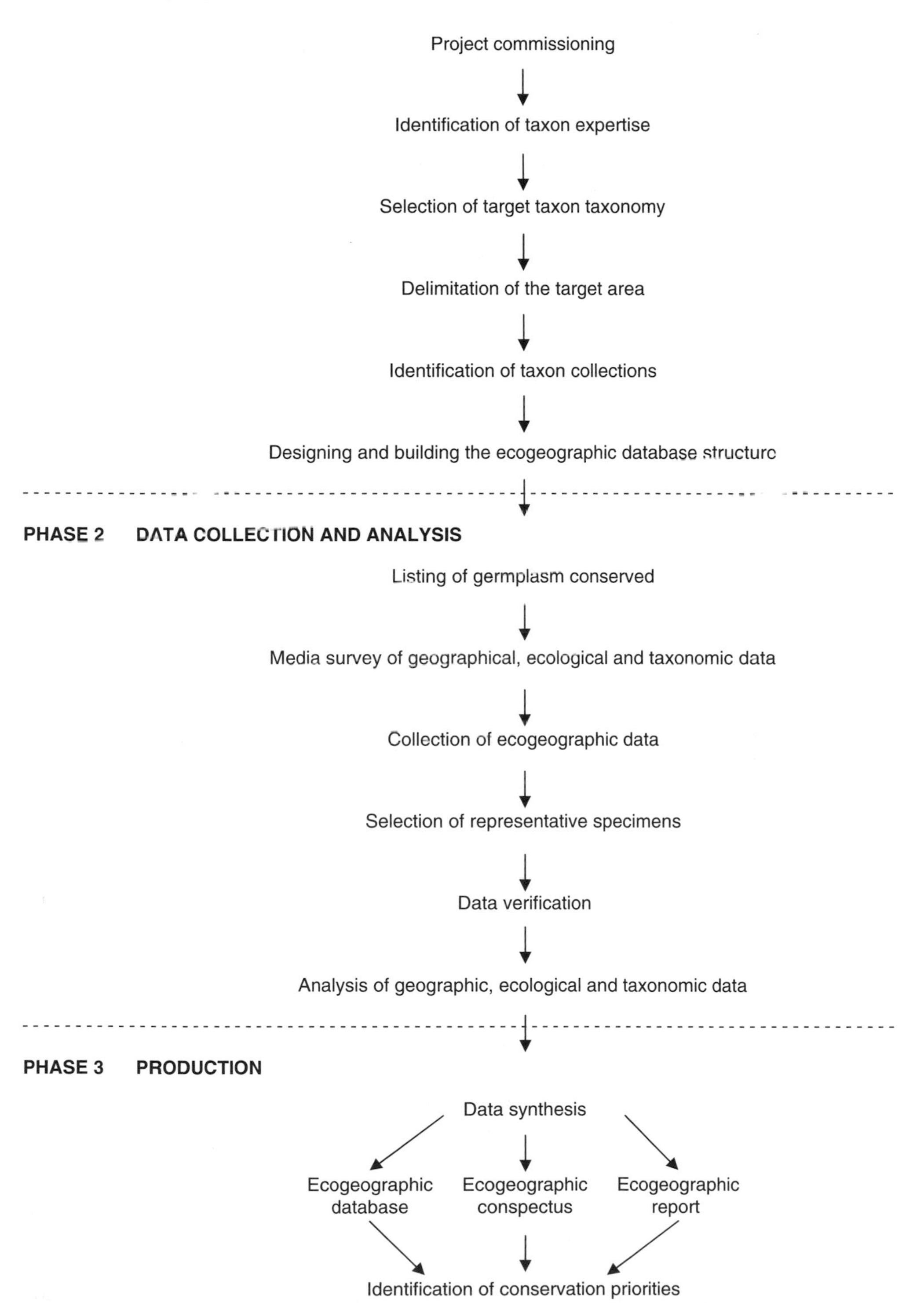

Fig. 2.1. Different steps for an ecogeographic survey or study. (From Maxted *et al.*, 1995.)

Local names of target taxon should be properly documented through surveys with local communities. The value of local names is that they allow comparison of target taxa between different sites and help ascertain whether one is dealing with the same or different entity. Although important, the existence of synonym and homonym in local plant names should also be considered. Heywood and Dulloo (2005) caution the use of common names to identify material. They are often locally specific but not unique over larger areas, and are often inaccurately associated with scientific names (Kanashiro *et al.*, 2002).

It is important to note the debate among modern botanists regarding the concept of species and the significant differences which exist between the different schools of thought. For example, for the same level of morphological differences, European botanists use the term 'subspecies', while in America, botanists prefer the use of the term 'variety'. Similarly, nowadays conservationists in Europe tend to use subspecies rather than species as the basic taxonomic level. In addition, it is important to consider the special cases of apomictic plants often called microspecies, which may have smaller but recognizable differences between them than is normal between species of most genera. The variability sometimes observed among different populations has also generated a vivid discussion between taxonomists. In some genera each micropopulation has been proposed as a different species, as in most Spanish species of *Alchemilla*. The number of newly recognized species, as a result of the updating and renewal of taxonomic concepts in the Standard Floras, is rapidly increasing. For instance, the flora of Spain was known to have about 5500 species of vascular plants, including both native and naturalized taxa, but as a result of the revision of the Flora Iberica and the revisions of the flora of the Canary Islands, this number can be expected to be over 7500. These taxonomic revisions have been made in many cases as the result of more accurate studies on the taxonomic variation observed between fragmented populations, where primitive species have been shown to evolve into two or more subspecies or even new species, due to adaptation to their microhabitats. In the case of some apparently common and well-distributed species like *Hedera helix* L. in eastern Spain, for example, it has been divided into two very apparently different subspecies: *H. helix helix* living in forests and rivers, and *H. helix rhizomatifera* mostly living as rupicolous plant (McAllister and Rutherford, 1990; Rutherford *et al.*, 1993; Vargas *et al.*, 1999). In locating genetic reserves, all the populations which in effect constitute the genetic diversity of the target species need to be taken into consideration.

2.2.2 Demographic information

Ecogeographic survey will identify broad areas where reserves could be established but they are unlikely to provide the best suites of specific locations representative of the diversity of the target taxon. Additional information is needed. In this respect, demographic information is very relevant in reserve selection as it includes the distribution range of the populations of the target taxon, numbers of populations, numbers of individuals within populations, and age-class distribution. The delimitation of the distribution range of the target species is critical in this exercise. Various types and sources of information can be used in defining the distribution range of a species. Passport data associated with herbarium specimens (in herbaria) and gene bank accessions (in gene banks) are often a first source of information for defining locations from where the accessions have been collected. This approach should be used with caution because it would provide only those areas where collections have been made and may not represent the entire distribution of the species. However, often herbaria are the only sources of information for determining geographical distribution of target taxa, especially non-crop species, for diversity and conservation studies (Maxted and Kell, 1998).

As mentioned previously, the unit of management in *in situ* conservation is often the population (see Heywood and Dulloo (2005) for a discussion). The number of populations across the distribution range should be determined and their characteristics studied and documented. This would include the number of individuals in each population and in particular the number of mature individuals, which according to the IUCN is the number of individuals known, estimated or inferred to be capable of reproduction (IUCN, 2001). This information is critical in determining effective population size (EPS) (see Iriondo *et al.*, Chapter 4, this volume). For most practical purposes, it is difficult to establish a minimum viable population (MVP), but it can be inferred from information obtained from demographic studies.

The age-class distributions within and across populations can be useful indicators of the status and viability of the population. The ratio of the various age groups in a population determines the current reproductive status of the population and indicates what may be expected in the future (Odum, 1971). For example, a rapidly expanding population will contain a large proportion of young individuals, a stationary population will have a more even distribution and a declining population will have a larger population of adults. Depending on the shape of the age pyramids, one can predict viability of the population. One should also consider disturbances in the context of their long-term impact on population persistence and health. For example, recent fire may have killed young individuals but the species is known to be well adapted and should regenerate in the next growing season. This reiterates the importance of understanding the biological and ecological attributes of the taxa.

Information on reproductive biology is another parameter of importance. For many species detailed information may not be available, but it is often possible to obtain information at the genus level, indicating if and when there are differences among species. There is a wealth of information on reproductive biology in standard texts on crop plants such as that of Purseglove (1968, 1972), and encyclopedias such as PROSEA publications (Plant Resources of South-east Asia). Simple web searches using Google will also reveal useful references and reference lists. Examples include NewCROP Purdue University web site (http://www.hort.purdue.edu/newcrop/default.html) and UNBC site for abiotic pollination (http://web.unbc.ca/bems/Abiotic-Pollination.htm).

2.2.3 Genetic variation

Knowledge of the amount and distribution of genetic diversity in target species is among the major criteria in locating sites for genetic reserves. In fact, the main objective for setting up a genetic reserve is to ensure that maximum genetic diversity of the target species gene pool is captured in the reserve system. This diversity is essential for species to evolve in their changing environment and to ensure their long-term persistence and survival (Frankel and Bennett, 1970). This requires that the level and patterns of genetic diversity both within and across populations are known (Neel and Cummings, 2003a). Genetic diversity occurs at various levels from the ecosystem, through its component species, populations, family groups, landraces and genotypes to the molecular level. Within each species there is likely to be substantial variation between individual plants (genotypes). At the gene level, allelic differences could be the basis of valuable traits of actual or potential value to changing environmental conditions and future use (resistance to pests, environmental stress, etc.). All levels of genetic diversity should be considered and, to the extent possible, included in the objectives of a conservation programme.

Depending on the distribution pattern of such variation, which in turn depends on the nature of the species' breeding/dispersal system, highly valuable genetic resources or hot spots of diversity can be located that can be targeted for protection within a genetic reserve. The amount of genetic diversity can most directly be assessed or measured by a range of molecular methods including, RFLP, RAPD, AFLP, microsatellites and SSR, and indirectly by biochemical (proteins, chemical markers) and morphometric characterization. This chapter is not intended to describe these various techniques, but the reader can refer to relevant publications such as Mallett (1996), Newbury and Ford-Lloyd (1997) and de Vicente (2004).

In practice, it is unlikely that information on the amount and distribution of genetic variability will be available for most species (Thomson *et al.*, 2001; Neel and Cummings, 2003a), and studies to generate this information can be too costly and time-consuming. In a review of the literature on this subject, it was found that there is no single framework for ensuring that maximum coverage of genetic variation in protected areas, although particular management recommendations have been published on this subject (Marshall and Brown, 1975; Soulé and Simberloff, 1986, Center for Plant Conservation, 1991; Lawrence, 2002; Neel and Cummings, 2003a,b). Maxted *et al.* (1997a) and Heywood and Dulloo (2005) discussed the problems linked to the lack of knowledge of genetic variation and their implication for *in situ* conservation and reviewed different alternative options to infer the expected distribution of genetic variation. Thomson *et al.* (2001) also provide guidance on what steps can be taken when there is insufficient data on genetic variation.

The following are the main proxy information that could be used to infer information on genetic variation and can help conservation planners in their task of capturing maximum genetic diversity for their target species:

- Ecogeographic representation. This involves choosing representative sites across the geographic range of the species in question (Maxted *et al.*, 1995; Thomson *et al.*, 2001). It assumes a direct relationship between genetic diversity, and geographical location is also represented.
- Genecological zonation (Graudal *et al.*, 1995, 1997; Theilade *et al.*, 2000, 2001). This can be a very practical tool to use in selecting populations when genetic studies are not available. It is based on the assumption that genetic variation follows some of the patterns of ecological variation (Thomson *et al.*,

2001). The technique involves defining genecological zones based on criteria of available genetic studies, local distribution of ecosystems, climatic information, physiographic maps, and geological and soil surveys. A number of populations within each genecological zone are then selected based on the biological characteristics of the target species, for example, their breeding system and level of endemicity, as well as on the level of risk or threat of their populations and resources available (Thomson *et al.*, 2001).
- Morphological characters of existing material maintained in *ex situ* conditions can play an important role in the analysis of variation in crop species and their relatives, largely because their collection does not require expensive technology (Newbury and Ford-Lloyd, 1997).

However, use of a proxy in the absence of such detailed knowledge of the patterns of genetic diversity should be handled with care (Maxted *et al.*, 1997b; Hawkes *et al.*, 2000). The assumption that sampling from geographically and ecologically diverse sites will provide the best sample of a species genetic diversity has been shown not to hold true for all species, as studies of wild lentils, for example, have proved. Ferguson *et al.* (1998) found that although wild lentils were distributed longitudinally from Portugal to Uzbekistan, the bulk of genetic diversity was concentrated within a very restricted segment of the geographic range of the taxon, namely the Fertile Crescent elements of Turkey, Syria and Iraq. The important point being the overriding value of genetic diversity information when available in locating sites to establish genetic reserves to ensure that the maximum genetic variation is captured, ecogeography provides a necessary proxy where such genetic diversity information is unavailable, but as a proxy it is fallible and detailed knowledge of the patterns of genetic diversity is always preferable.

The question of how many populations are necessary to capture maximum genetic diversity is an important one, and one on which there has been much debate (Marshall and Brown, 1975, Brown and Briggs, 1991, Center for Plant Conservation, 1991; Lawrence, 2002; Neel and Cummings, 2003a,b). These authors provide guidance as to the number of populations and individuals that need to be sampled to ensure the *ex situ* conservation of a certain level of genetic variation. Commonly cited recommendations include:

- Sampling from five populations to effectively capture 90–95% common alleles (Brown and Briggs, 1991);
- Collecting 50 individuals from 50 populations (Brown and Marshall, 1995).

Neel and Cummings (2003b), however, demonstrated that five populations would conserve on average 67–85% and as little as 54% of all alleles if populations were selected randomly without knowledge of genetic diversity patterns. Brown and Schoen (1992) also mention that in outbreeding plants the average single populations appear to have over 80% of the species genetic variation as measured by the gene diversity index. If this is true, the conservation of very few populations of each species would be adequate for future conservation (Brown and Schoen, 1992). Lawrence *et al.* (1995) also recommended that a sample of 172 plants drawn at random from the populations of a target species is sufficient to capture at a very high probability most of the diversity in the populations of that species. Neel and Cummings (2003a) investigated the genetic consequences of ecological reserve guidelines and made the following important conclusion:

> [S]electing populations according to common ecological reserve guidelines did not capture more genetic diversity than selecting populations at random. The number of populations selected is much more important than how those populations were selected. Thus, focusing on ecological features for selecting sites for conservation will ensure representation of genetic diversity only when sufficient numbers of populations are included in reserves.

There cannot be one solution to this problem. For most practical purposes in setting up a network of genetic reserves, the conservation planner would need to call on his or her best judgement depending on such factors as social, economic and political circumstances. But based on the principle set by Brown and Briggs (1991) it would be recommended to select a minimum of five populations for each species.

2.2.4 Ecological information

Information on the target species' environment is just as important as information on the target species itself to ensure the long-term persistence of the target species in its natural habitat. The different types of information under this category would be very helpful in the selection of sites for conservation:

Ecosystem condition and function – This information is important because ecosystem condition, quality of ecosystem services and ecosystem resilience indicate the system's capacity to maintain component species and in particular the target species.

Threats to the habitat – It is also important to identify not only what threatens habitat condition, but also the degree and extent of threat, which would influence the decision of conservation planners in identifying sites for inclusion in the reserve system. It also has direct implications for the cost of reserve management in terms of threat mitigation. Threats include invasive species that alter vegetation composition and ecosystem dynamics, frequent and intense fire, overgrazing, insect pests and pathogens, pollution and climate change (see Section 2.3 on current conservation status). Ecosystems are subject to natural and human-induced disturbances and threats at various spatial and temporal scales. In the long term, it is difficult to exclude large disturbances. Indeed, many CWR species, such as rice and lettuce, persist in disturbed or semi-disturbed sites. Reserves should not be isolated and static, but be regarded as part of dynamic landscapes (Bengtsson *et al.*, 2003).

Climate change – This is a significant threat to the maintenance of biological diversity and ecological systems. It is clear that the relatively modest climatic changes over the last century have had a significant influence on the distribution, abundance, phenology and physiology of a wide range of species. Many shifts of species distribution towards the poles or upwards in altitude have been recorded, with progressively earlier migrations and breeding (e.g. Parmesan and Yohe, 2003). While many populations of CWR species will be adversely impacted by climate change, conservation actions targeting this group of economically important plants are rare. In a recent study on wild *Arachis* (groundnut), wild *Solanum* (potato) and wild *Vigna* (cowpea) to predict current and future species distribution,

Jarvis *et al.* (in press) predicted that by 2055, 24–31 of 51 *Arachis* species, 8–13 of 107 *Solanum* species and 1–3 of 48 *Vigna* species could become extinct. These results are particularly interesting because they indicate the magnitude of threat and also the differential impact from one crop complex to another. The study also projected that habitat patches will become more fragmented, exacerbating vulnerability to further erosion.

Alteration of ecosystem composition, structure and dynamics as a result of climate change must be considered when developing criteria for selecting new protected areas and managing existing ones. The combined effects of habitat fragmentation and climate change are serious issues for biodiversity conservation (Young and Clarke, 2000). Fragmentation creates barriers to migration and recolonization and as habitat areas decrease, they become increasingly vulnerable to suboptimal population sizes and further erosion. It is likely that climate change will result in species being lost from protected areas. Araújo *et al.* (2004) projected that 6–11% of 1200 plants with European distribution could be lost from reserves in a 50-year period. Thus, long-term conservation of species within protected area boundaries may be aimed at a moving target (Van Jaarsveld *et al.*, 2003).

For species with overlapping distributions between current and future climate regimes, reserve networks could be identified to represent both current and future-projected occurrences; for species with no overlap in distributions, new areas must be identified. These might include areas where the impacts of climate change are expected to be minimal and areas where new climates can provide habitats for emigrating species. Given that plant populations tend to migrate to higher altitudes (Parmesan and Yohe, 2003), establishing protected areas in mountainous zones may be a priority. Areas selected for conservation should be evaluated for their potential as climate refugia for vulnerable species, sources of propagules for recolonization and of genes for *ex situ* collections.

Habitat diversity – When selecting sites for establishing a genetic reserve, sites with spatial or temporal heterogeneity should be given priority over homogeneous areas. In general, the wider the range of habitat diversity and juxtaposition of different habitats within a potential site, the better the reserve will be at conserving maximum diversity, both at intraspecific and interspecific levels. This will ensure that the target species will preserve the various genes and genetic combinations associated with ecotypic and taxonomic differentiation leading to more effective conservation of genetic diversity. Should the target species be restricted to a few soil types, certain levels of soil humidity, or certain natural or semi-natural habitats, these must obviously be taken into account when the reserve is being planned and appropriate areas or habitats included (Yahner, 1988). This has led to theories of landscape ecology (Forman, 1987; Urban *et al.*, 1987; Hansson *et al.*, 1995). It is also important to consider a diversity of microhabitats with the network of genetic reserves. A representative case applied to CWR is the model of 'Plant Micro-reserves' (PMRs) (Laguna, 2001, 2005; Laguna *et al.*, 2004), where the reserve location not only is dependent on the species information, but also considers in all small sites holding microhabitats rich in endemic, stenoic species such as rock crevices, temporary ponds, etc. in the PMRs network of the Valencian Community, Spain (Plate 3).

Care should also be taken when applying management regimes to genetic reserves; when a particular site is selected and a management regime instigated, the habitat may in its early stages lose diversity. Therefore, the management regime may necessarily include habitat disturbance, which results in the desired patchwork of diverse habitat types. Natural causes of disturbance include fires, storm damage, pest and disease epidemics, herbivory, floods and droughts. All of these factors are non-uniform, in terms of coverage, and create habitat patches of an earlier successional stage, which will in turn promote species and genetic diversity. Pickett and Thompson (1978) discuss the selection of reserve sites in conjunction with habitat disturbance and heterogeneity. They refer to minimum dynamic area, which is the smallest area with a complete, natural disturbance regime. This area would maintain internal recolonization to balance natural extinctions. As the majority of species are not exclusive to one habitat, the maintenance of reserve heterogeneity will promote the overall health (genetic diversity) of the full gene pool as represented in multiple populations or metapopulations.

The need for continued habitat heterogeneity is a factor that will need to be considered when formulating the reserve management plan (see Maxted *et al.*, Chapter 3, this volume). For example, if fire were a natural causal agent of habitat disturbance and heterogeneity which promoted the target taxa, the reserve design would have to permit continued use of fire, under the instigation of the reserve manager. If a reserve site is designated that includes human habitation, regular fires may be undesirable or even dangerous. Also, information on the status of associated species, in particular the presence or absence of known pollinators, dispersers and any symbionts for the target species, as well as their population levels, should also be taken into account. Wherever possible, reserve sites should be linked through habitat corridors or stepping stones to facilitate the appropriate movement of associated pollinators or dispersers. In practice, areas with higher species richness will receive more attention, and species richness is, in fact, a major criterion used in reserve selection. So the ultimate management regime will often affect the reserve design.

Species distribution – The geographical distribution of target species can be predicted using Geographical Information Systems (GIS) (Jones *et al.*, 1997). Several commercial GIS software (such as ARC-INFO, ARC-VIEW, ARC-GIS, IDRISI) are available whereas other packages have been specifically developed for genetic resources work. For example, FLORAMAP was developed at CIAT (Jones and Gladkov, 1999; Jones *et al.*, 2002) and DIVA-GIS was developed jointly between CIP and Bioversity (Hijmans *et al.*, 2001). A list of references on spatial analysis and GIS applied to genetic resources management is available on the Bioversity web site (2007). In GIS work, different parameters, usually climatic data such as temperature and rainfall, as well as soil data, are used to determine potential locations. Also, the preferred habitat type of the target species may be a useful parameter to include in the analysis. Habitat distribution can be helpful in predicting suitable locations for the taxon or to identify sites for recovery programmes (see Kell *et al.*, Chapter 5, this volume). Sites identified using the above methods should always be validated in the field to assess the presence or absence of the target species and the state of the habitat.

Predicting species ranges for different climates is commonly done using climate envelope models (see Box 2.1) which use a species' environmental requirements known from localities where the species currently occurs (Hijmans and Graham,

Box 2.1. Species distribution modelling in the face of climate change.

As seen in this chapter, the composition and genetic diversity of species in a given reserve is not static in time. The current threat of climate change may create a drastic shift in species distribution, and so areas targeted under current conditions for conservation may not necessarily conserve the same biological resources in the future. This makes the targeting and conservation for genetic reserves more complex, especially when the reserves are identified with specific species in mind. Basic data about species distribution therefore become a key information resource for the planning of genetic reserves, and through modelling, species distribution can be predicted under current and future climatic conditions.

Species distribution modelling: Much effort has gone into the development of methods for predicting the geographic distribution of species and now many of these have been incorporated into user-friendly tools. Typically, these methods use the conditions at points where the species has been found in order to construct a statistical model of the adaptation range of the species, based on a set of user-defined environmental variables. The statistical model then is applied over a wide region to locate other areas where the environmental conditions are potentially suitable for the species in question. Climate is one of the major factors governing the distribution of wild plant species, acting directly through physiological constraints on growth and reproduction (Walker and Cocks, 1991; Franklin, 1995; Guisan and Zimmerman, 2000) or indirectly through ecological factors such as competition for resources (Shao and Halpin, 1995). When a species distribution is predicted using climate variables only, it is commonly referred to as a climate envelope model.

Climate envelope modelling applied to climate change: Global warming has accelerated over the last 30 years (Osborn and Briffa, 2005), and is predicted to be in the range of 1.1–6.4°C by 2100 (IPCC, 2007). Modelling studies (e.g. Thomas *et al.*, 2004) indicate that climate change may lead to large-scale extinctions. A number of studies have applied climate envelope-based species distribution models to the problem of understanding the impacts of climatic change through the use of climatic data for the present and the future (Thomas *et al.*, 2004) and the past (Ruegg *et al.*, 2006). These methods essentially transfer a species adaptation temporally, assuming on the one hand no more plasticity than is currently observed and on the other zero evolution, and many overlook the possible consequence of changes in biotic interactions such as competition (Lawler *et al.*, 2006). There is a growing body of research evaluating the suitability of applying species distribution models to predicting range shifts and assessing extinction risk in the face of climate change (Thuiller *et al.*, 2004; Araújo *et al.*, 2005a,b; Araújo and Rahbek, 2006; Hijmans and Graham, 2006; Lawler *et al.*, 2006). Despite concerns about the difficulty of validating such modelling studies (Araújo and Rahbek, 2006), the results can provide useful insights of relevance to conservation planning as long as they are interpreted within the context of the assumptions and uncertainties made in the modelling.

Sources of climatic data for climate envelope modelling: There is a great deal of climate information available freely for any part of the world. The most comprehensive data set is the WorldClim database (Hijmans *et al.*, 2005), consisting of global surfaces of mean, monthly mean, maximum and minimum temperatures and rainfall for the period 1960–1990 with a spatial resolution of just 1 km. The data are available freely on the Internet (http://www.worldclim.org), along with a set of 19 bioclimatic variables derived from monthly means which are of use in climate envelope models for predicting species distribution. For future climates, General Circulation Models (GCMs) form the basis of predictions. A wide range

Continued

Box 2.1. *Continued*

of models and emission scenarios are available, typically at spatial resolutions too coarse for species distribution modelling (>1°). However, downscaled results of some models (HADCM, CCM3 and CSIRO) and the most popular scenarios (a2a, b2a) are available from WorldClim (http://www.worldclim.org/futdown.htm). More detailed discussion of the different GCM outputs and their respective levels of uncertainty is available in the IPCC 4th Assessment reports (IPCC, 2007).

Climate envelope models: A wide range of climate envelope models exist that use different statistical approaches, including principal components analysis (Jones *et al.*, 1997; Robertson *et al.*, 2001), generalized linear models (Cumming, 2000; Pearce and Ferrier, 2000; Guisan *et al.*, 2002; Osborne and Suárez-Seoane, 2002; Draper *et al.*, 2003), factor analysis (Hirzel *et al.*, 2002), genetic algorithm (Anderson *et al.*, 2002) and maximum entropy (Phillips *et al.*, 2006). A comprehensive assessment of different climate envelope models is made by Elith *et al.* (2006), but some of the better and more user-friendly tools include:

MaxEnt – http://www.cs.cmu.edu/~aberger/maxent.html
Desktop GARP – http://nhm.ku.edu/desktopgarp/index.html
Bioclim – embedded in http://www.diva-gis.org

Implications for conservation: The application of climate envelope models to understand the shifts in species distribution under future climate change can inform conservation planning by identifying the areas where *in situ* conservation efforts should be focused, while identifying populations where *ex situ* conservation may be the most rational strategy for ensuring conservation of the genetic resources. Areas identified as potential refugia for a given species are clearly *in situ* conservation priority regions, while regions where species are predicted to be lost may not present the most sustainable places for reserve establishment, and *ex situ* conservation efforts may be required.

Case study example: Jarvis *et al.* (in press) applied climate envelope models to understand the shifts in species distribution of selected species of CWR under future climate change. Taking one important CWR of groundnut as an example, *Arachis batizocoi* is an economically important species closely related to the cultivar which has been widely used in breeding (Simpson and Starr, 2001). This species is currently predicted to cover a range of just over 16,000 km^2 in the Chaco region of South America. Applying climate envelope models to the future predicted climate under the CCM3 model and a2a emission scenario (representing a peak of 600 ppm CO_2 content in the atmosphere), this distribution area is predicted to be reduced to just 717 km^2 assuming that the species can migrate freely over large distances, or to just 261 km^2 if the species is unable to migrate spatially. This latter area represents the refugia for this species, and could be considered a priority region for a genetic reserve. Combining the analysis with other species might identify this area as refugia for a number of other important species.

2006). To improve the accuracy of models in predicting the status of species and populations, models also need to be coupled with land-use projection models, which represent the current pattern of habitat fragmentation and model future patterns based on projections of parameters such as population and consumption levels (Sala *et al.*, 2000). The potential range shift of a species approximated by the bioclimatic models is then reduced to the available habitat as projected by the land-use model.

Species migration – A central question and a major uncertainty in the application of species distribution models to understanding the impacts of climate change relate to the migrational (dispersal) capacities of species (Pearson, 2006). Species capable of migrating at high rates are more likely to survive, and indeed in some cases may gain geographic range, thanks to greater land mass in higher latitudes and to species–energy relationships (Menendez *et al.*, 2006). Species with the greatest mobility (plants with wind-dispersed seeds) will be the most capable of shifting with climate zones and should therefore indicate changes sooner than species with limited dispersal capacity.

Most modelling studies account for migration by assuming it to be either unlimited or non-existent, yet the reality is likely to be somewhere in between (Pearson, 2006). Some fundamental aspects of migration can be used to broadly estimate migration capacity. The mode of dispersal will have a direct influence on a species' ability to migrate. The most common modes of long-distance dispersal for terrestrial plants are by wind, water, animals and humans. For wind dispersal, the distance of transport can be estimated using information on the direction of predominant winds coupled with prevailing weather conditions. Water dispersal can also be predicted by direction and speed of flow. However, seed survival in water is an important factor. For animal dispersal, transportation is less predictable and will depend on the species of animal and the morphological characteristics of the seed (Neilson *et al.*, 2005). Humans transport seeds and propagules knowingly or unknowingly in vehicles, often in topsoil and agricultural products. Invasive species are often spread in this way (Pysek *et al.*, 2002).

There is currently no analysis of what proportion of local, regional or global floras may persist or adapt to climate change *in situ*, but it is widely recognized that migration at a rate that keeps pace with climate change is a necessary response in organisms that lack extreme stress tolerance and longevity of genetic plasticity (Midgley *et al.*, 2007).

Phenotypic plasticity – The capacity of individual species to adapt to changing conditions may well be an important determinant of population persistence in the face of climate change. Phenotypic plasticity is the environmentally driven trait expression within a genotype and lies at the heart of species evolution (Valladares *et al.*, 2006). It has frequently been reported as the primary mechanism enabling exotics to colonize environmentally diverse areas (e.g. Williams *et al.*, 1995; Niinemets *et al.*, 2003; Peperkorn *et al.*, 2005) and may well be an important enabling process for persistence of populations under changed climatic conditions. Projections of species distribution using bioclimate envelope models may overestimate species losses if plasticity is ignored (Thuiller *et al.*, 2005). In selecting potential sites for genetic reserve the variability offered by phenotypic plasticity should be an important criterion particularly in the context of climate change. In this respect, the apparent lack of some phenological behaviour (e.g. flowering, seed reproduction) at given times should not be construed as an 'abnormal' behaviour and a possible reason to reject a site as potential reserve location. For instance, Laguna (Valencia, Spain, 2007, personal communication) has observed that there exists tremendous variation in the cycle of significant seed production for some tree species such as *Quercus faginea* across Spain, which happens every 20–25 years in eastern Spain but the same species show much shorter cycles in central and western Spain.

2.2.5 Policy and socio-economic information

In addition to biological factors, policy, socio-economic and cultural factors of the target taxa should also be considered in locating potential genetic reserve sites. Not only are these factors likely to influence the choice of sites, but they are also likely to determine the kind of interventions necessary to sustainably maintain populations of the target species *in situ*. It is recommended to carry out ethnobotanical surveys to document the use and level of harvesting of target species from natural habitats. Local communities are also able to provide a wealth of information on the location of target species in their areas as well as the traditional knowledge associated with those species (Plate 4).

Legal status – It is always useful to determine the historical land tenure of the potential sites and the current ownership status of the land. This will have major legal implications in the management of the site and in particular with regard to accessibility and land rights as *in situ* genetic conservation involves the permanent appropriation of land and the management of the sites, thus raising a series of constraints ranging from economic to scientific and organizational matters (Williams, 1997). The legal status of the species and the possible sites is also an important factor to be taken into account. In the case of protected species, its status can vary greatly between different countries and regions (de Klemm, 1990, 1996, 1997; de Klemm and Shine, 1993) and will influence interventions that can be made to ensure their conservation. For instance, if a taxon is not formerly included in the official recovery plans which, for some countries like Spain, have legal status passed by regional decrees, there can be severe restrictions to the kind of interventions permitted. Reciprocally, the design of recovery plans for plant conservation should also include the proposal of reserve locations (Wyse Jackson and Akeroyd, 1994).

The legal status of the site should also be analysed in the context of what kind of protection the legal framework provides. The models of legal framework can be extremely diversified. Laguna (2001) reviewed the legal and technical models of small protected sites and reserve networks for wild plants around Europe, the former Soviet Union, the Middle East and North Africa, and showed that some apparently single, very local models, can fit the protection of rich biodiversity areas, as in the case developed by the Marche region in Italy (de Klemm, 1997). The monographic works of de Klemm (1990, 1996, 1997) also give an accurate overview of the larger topic of plant species, habitats and sites protection. It should be noted that the legislations quickly changed country by country during the past years, due to effective political changes (i.e. incorporation of new members to the European Union, forced to adopt the EU rules and directives), and it is thus important to keep updating the information from the web sites of the governmental institutions caring for nature conservation at national, regional and local levels. In addition, the long-term management can be ensured and reinforced by means of the active participation of conservationist NGOs, custodian agreements with landowners and other strategies (Shine, 1996).

In many cases, legally protected areas give a blanket protection against site destruction and do not allow any kind of management. A classic case is the UK Sites of Special Scientific Interest (SSSI) (Laguna, 2001). Legally protecting areas is not sufficient and active management of the site may be required for the safeguard of the target species. Securing a long-term lease from government authorities or landowners for managing protected areas often helps achieve the conservation potential of the sites. In Mauritius, for example, a small islet nature reserve, Ile aux Aigrettes, was highly threatened until a long-term lease provided by the government in 1984 to a local NGO, Mauritian Wildlife Foundation, allowed a restoration plan to be developed and implemented (Dulloo *et al.*, 1997). On the other hand, many well-managed European sites, acting as effective reserves, are not legally protected. For example, in Belgium many conservationist NGO sites have custodianship agreements with landowners, and local governments manage the sites to ensure effective conservation of the biological assets of the site. In addition, many protected sites around the world have been set aside to conserve charismatic fauna such as birds, large mammals and so on, and the management of such reserves often (though not always) conflicts with plant species. For instance, the conservation of the endemic sea-lavender plant *Limonium dufourii* in Valencia (Spain) depends on the agreement between the regional governments which partially share their distribution area and habitats (salt scrublands and grasslands) with one of the most endangered animal species of Europe, the marbled teal *Marmaronetta angustirostris*. Also, the marbled duck's population is controlled by artificially inundating the level of freshwater in the same habitat where *L. dufourii* lives, thereby causing abnormal changes in its population dynamics (Navarro *et al.*, 2006).

In some EU countries (Lithuania, Hungary, Slovakia) and elsewhere there are approved national legal acts defining, *inter alia*, genetic resources conservation *in situ*. In Lithuania two national laws related with genetic reserves were ratified by the Parliament in 2001: Law on National Plant Genetic Resources (http://www3.lrs.lt/cgi-bin/preps2?Condition1=230551&Condition2=) and Law on Protected Areas, which specifically includes micro-reserve. In Hungary the Ministry of Agriculture adopted a Decree on PGR in 1997 (http://www.rcat.hu/english/activity/decree/decree.htm). These kinds of acts make a legal background to the establishment of genetic reserves.

Economic considerations – While the biological and ecological aspects of reserve selection have been well studied, relatively little attention has been paid to the economic aspects of biodiversity conservation (Naidoo and Ricketts, 2006). The cost of reserve management is a critical factor in site selection. Resources need to be strategically allocated to maximize conservation gains from the limited resources available. It is highly recommended to perform a cost–benefit analysis for informing conservation decisions. In general, the costs include land prices, management costs – which will vary according to the type of interventions required (see Maxted *et al.*, Chapter 3, this volume) – and opportunity costs. These costs should be compared with the benefits afforded through the conservation of the resources along with the value of CWR themselves as well as the ecosystem services they provide, including carbon sequestration and storage, timber harvesting and flood control (Naidoo and Ricketts, 2006). The cost of establishing and managing genetic reserves varies substantially between countries (Balmford *et al.*, 2003), depending on land prices and benefits foregone and generated. There are many challenges in quantifying both costs and benefits, especially in valuing environmental services, such as those derived from CWR, on a spatial scale (Balvanera *et al.*, 2001; Turner *et al.*, 2003). Further, the perceived value of services may vary between beneficiaries.

All of the information gathered above should then be analysed and the results fed into the next step of site selection and design, where criteria for site selection are discussed.

2.3 Reserve Site Selection for Single Target Taxa

The amount or quality of information obtained may not always allow reserve sites to be systematically selected. Also for economic, social or other valid reasons, it may not be possible to conserve all the sites where the target species occur. Therefore, prioritization of conservation areas is a must and forms an essential part of biodiversity conservation (Noss and Harris, 1986; Margules and Pressey, 2000; Kjaer *et al.*, 2004). The choice of precise sites for the conservation of target species involves setting goals, targets and scales (Balmford, 2002). In practice, conservation planners will have to use as much relevant information as is available and apply selection criteria to prioritize sites.

The selection criteria should be based on the goal or objective of the conservation programme. According to Soulé and Simberloff (1986), these criteria can be biological as well as cultural, political or economic. Other authors have indicated different criteria of selection (Margules and Pressey, 2000; Kjaer *et al.*, 2004). However, it is important to recognize that different criteria would be applied to different situations depending on the objective of the conservation programme, and there can be a mix and match of these criteria. Most of the conventional criteria used for priority setting include the identification of species-rich areas or hot spots (Williams *et al.*, 1996), or using vegetation communities or assemblage (Mackensie *et al.*, 1989). The focus can also be on perceived threats resulting from economic value or ecological traits (Kjaer *et al.*, 2004). Genetic data are seldom used in prioritizing conservation areas and only a few examples exist. One example is in the Western Ghats of India, where molecular techniques have been used to detect areas of high intraspecific and interspecific diversity to set priorities for conservation (Boffa, 2000). Another example is the Gene Management Zone (GMZ) set up in Turkey to conserve genetic diversity in wild species *in situ* (Tan and Tan, 2002). Australian conservationists proposed the so-called representation approach (Pressey *et al.*, 1997) which could be employed in the selection of different land types and better representation of target populations.

Pressey *et al.* (1997) identify four representation problems for which solutions might be required:

1. The minimum number of sites needed to represent at least one occurrence of each feature;
2. The minimum total area of sites needed to represent at least one occurrence of each feature;
3. The minimum number of sites needed to represent at least 5% of the total regional extent of each feature;

4. The minimum total area of sites needed to represent at least 5% of the total regional extent of each feature.

In the test application they described, Pressey *et al.* (1997) sought to ensure representation of 248 land systems across 1885 potential conservation sites in an area of 325,000 km^2. They found that a minimum of 54 sites (2.86% of the total) and an area of 12,084 km^2 (3.72% of the total) are needed to represent at least one occurrence of each land system. They also found that a minimum of 126 sites (6.68% of the total) and an area of 25,887 km^2 (7.96% of the total) are needed to represent at least 5% of each land system.

In locating genetic reserves, the following criteria should be considered:

- *Level and pattern of genetic diversity of the target species' populations*: When a characterization of genetic diversity has been made in a representative sample obtained throughout the taxon's distribution area, the set of populations that, according to the results, maximize within-population as well as between-population genetic diversity should be chosen. When no previous genetic studies are available, ecogeographic or genecological proxies are used.
- *Presence in protected areas or centres of plant diversity, or centres of crop origins or diversification*: Priority should be given to sites already present in protected areas. Margules and Pressey (2000) recommend to first review the extent to which target species are well maintained (in terms of representation and persistence) in existing reserves. They recommend making use of gap analysis to identify any gaps in the network of already extant reserve networks. Only after this stage can the need to include other areas be assessed. Many existing protected areas have operational management plans. Including specific management actions for target taxa into an existing plan is likely to be more cost-effective than establishing a new protected area and developing a management plan for the entire area.
- *Size of reserves*: The size of the reserve will depend on the characteristics of the target species and its symbiotic relationships. This criterion is also of importance in reserve design (see Section 2.5).
- *Number of populations*: The number of populations and individuals is often so reduced that there are no options other than to try and save what is available rather than any theoretically recommended MVP. Populations which contain important genetic, chemical or phenotypic variants should receive a higher consideration. In target species with a larger number of populations a minimum of five populations should be chosen for the genetic reserve network of a particular CWR species.
- *Number of individuals within the population*: A minimum number of viable individuals is needed to maintain genetic diversity. Considerations of MVP are discussed in Section 2.5.
- *Political and socio-economic factors*: The location and design of any reserve is rarely decided solely on the basis of biological considerations; political and economic factors often play a decisive part. The needs of local communities must be taken into account in the selection of reserve sites. Areas to be protected need to be defined in consultation with local people and a stakeholder analysis is recommended (Isager *et al.*, 2004). Declaration of genetic reserves could have implications for neighbouring communities whose use of the area for goods

and services may be impacted. Costs and benefits foregone, therefore, need to be estimated, as discussed earlier.

- *Current conservation status*: Current conservation status of species and their populations may be used as an indicator for prioritization (Graudal *et al.*, 2004). The degree of current and potential future threat, including impact of climate change on the population, should be a decisive criterion particularly when the population contains important genetic, chemical or phenotypic variants (Heywood and Dulloo, 2005). In the case of narrow endemics, they are most likely to be contained in a few, but critical, populations and should receive high priority. For more widely dispersed species, other selection criteria will need to be applied to locate numbers that are most cost-effective to maintain.

2.4 Reserve Site Selection for Multiple Target Taxa

As discussed above, practically it would rarely be the case that a genetic reserve is established for a single target taxon; it is more likely for economic reasons that reserves are established to conserve multiple target taxa as part of a national CWR conservation strategy. Within such a strategy *in situ* CWR conservation is preferable because of local ecotypic adaptation of genetic diversity and the need to conserve the full range of intraspecific diversity, and the sheer numbers of potential CWR involved, which both mean that *ex situ* conservation is impractical for all CWR except the most threatened or precious. If it is agreed that five reserve locations would be required to adequately conserve the genetic diversity of each species, implementing the national CWR conservation strategy would involve the establishment of networks of reserves for the long-term maintenance of biodiversity (Hopkinson *et al.*, 2000; Margules and Pressey, 2000; Ortega-Huerta and Peterson, 2004). The creation of such a network of national CWR genetic reserves is therefore likely to be a priority within any national CWR conservation strategy, as recommended by the Global Strategy for CWR Conservation and Use which is in preparation by the Food and Agriculture Organization of the United Nations (FAO).

The identification of the most appropriate sites for multiple target taxa genetic reserves may be viewed as a four-stage process (Maxted *et al.*, 2007): (i) creating the national CWR inventory; (ii) prioritizing the most important CWR taxa for active conservation; (iii) using distributional data for the CWR species circumscribed to identify national complementary CWR hot spots; and then (iv) matching CWR hot spots to the existing protected area network to identify existing protected areas where genetic reserves could be established (see Box 2.2).

Box 2.2 summarizes a detailed worked example of how a country might implement a national CWR conservation strategy using as an example the UK – but how typical is the UK? It is possibly true that the UK is atypical, in that its flora is extremely well studied, in many cases down to population levels for species, and the national network of protected areas is also systematic and comprehensive. As such, individual CWR species have been mapped at least to 10 × 10 km tetrad levels and it is possible to relatively easily match these distributions to the existing protected

Box 2.2. Establishing multiple CWR taxon genetic reserves: a case study from the UK. (From Maxted *et al.*, 2007.)

Creation of the UK national CWR inventory: The UK National Inventory of CWR was derived from the PGR Forum Crop Wild Relative Catalogue for Europe and the Mediterranean (Kell *et al.*, 2007) (www.pgrforum.org), which in turn was generated by data harmonization and cross-checking between a number of existing databases, primarily Euro+Med PlantBase (http://www.euromed.org.uk), Mansfeld's World Database of Agricultural and Horticultural Crops (Hanelt and IPK 2001; http://Mansfeld.ipk-gatersleben.de/Mansfeld/), with forestry genera from enumeration of cultivated forest plant species (Schultze-Motel, 1966), ornamental genera from the Community Plant Variety Office (www.cpvo.eu.int), and medicinal and aromatic plant genera from the MAPROW (Medicinal and Aromatic Plant Resources of the World) database (U. Shipmann, Bonn, Germany, 2004, personal communication). The resulting UK National Inventory of CWR database can be queried via the UK Genetic Resources for Food and Agriculture portal (http://grfa.org.uk/search/plants/index.html?#sr). The database contains basic taxonomic and usage data, along with conservation data such as occurrences and trends, legal status, IUCN threat assessment status and conservation action plans. Due to inconsistencies between European and UK plant nomenclature, once the initial UK catalogue was extracted from the European and Mediterranean catalogue, it was necessary to standardize the nomenclature to that applied within the UK using the standard national flora, *New Flora of the British Isles*, second edition (Stace, 1997). In addition to the data extracted from the European CWR inventory, UK data sets were included, such as use categories, occurrences and trends, UK legal status, IUCN red list assessment and whether conservation action plans were available. The UK National CWR Inventory contains 113 genera and 1955 species (2644 if microspecies and subspecies are included), which means that 65% of UK native species are CWR (Maxted *et al.*, 2007).

Prioritizing UK CWR taxa: The sheer number of UK CWR species means that prioritization is necessary. Although many criteria can be used to establish priority taxa for conservation (Margules and Usher, 1981; Maxted *et al.*, 1997c; Ford-Lloyd *et al.*, 2007), relative threat assessment using the IUCN Red List Criteria and economic value were used to prioritize UK CWR taxa, and this generated a priority list of 250 UK CWR species.

Identification of UK CWR hot spots: Detailed UK distribution data for 226 of the 250 CWR priority species were obtained from the Botanical Society of the British Isles via the NBN Gateway (www.searchnbn.net) and overlaid to identify UK CWR hot spots. The initial task was to identify the 'best' sites in the UK with the highest CWR species coverage in which to establish genetic reserves. This was achieved using the iterative selection procedure that resulted in complementary conservation of maximum species diversity (Kirkpatrick and Harwood, 1983; Pressey and Nicholls, 1989; Pressey *et al.*, 1993; Rebelo, 1994; Bonn *et al.*, 2002). The site with the highest species number is allocated as the first site, then the species located in this first site are excluded from analysis and the second site is selected using the remaining species and so on (Rebelo, 1994). The question then arises as to how many CWR taxa need to be present to be regarded as a hot spot and how many hot spots can pragmatically be nominated to effectively conserve CWR diversity. The

Continued

Box 2.2. *Continued*

data illustrate that as the number of 10 × 10 km tetrads increases, the percentage of diversity added decreases and the economic cost of adding sites for smaller diversity gain may therefore become less attractive (Fig. 2.2). If the aim is to conserve two-thirds of the total priority CWR species diversity, 17 tetrads would be required for the location of UK CWR genetic reserves. However, as can be seen in Fig. 2.2, the percentage of CWR diversity added levels off to less than 2% after the tenth site, and it was thought that recommending the establishment of 17 genetic reserves for CWR might not be accepted by protected area network managers; therefore, pragmatically it was concluded that the top 10 of the 17 should be recommended for the establishment of genetic reserves for UK CWR taxa (Fig. 2.3).

Matching UK CWR hot spots with existing protected areas: Once the UK CWR hot spots (10 × 10 km tetrads) were identified, they could be matched against the existing protected area network to identify potential sites where genetic reserves for *in situ* conservation of CWR could be established. The existing protected areas were Special Areas of Conservation (SAC) and Sites of Special Scientific Interest (SSSI) and these were compared with ten top CWR hot spots. It should be stressed that matching CWR hot spots with existing protected areas can only be used to predict CWR presence in protected areas; in each case, field visits would be required to confirm the prediction before the final site is selected for CWR genetic reserve establishment. Following the checking of species lists for protected areas within the ten tetrads it was found that they contain multiple populations of 128 (57%) of the 226 priority UK CWR species.

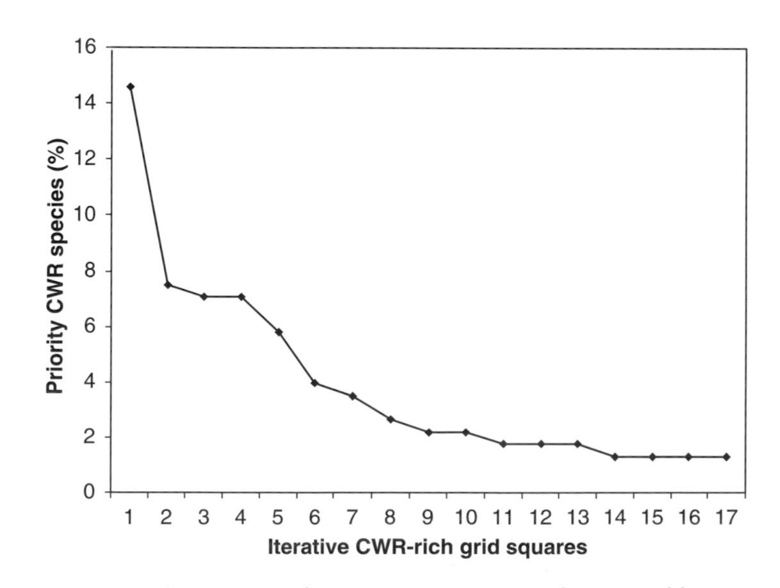

Fig. 2.2. Percentage CWR gain with increasing iterative grid square addition.

Fig. 2.3. Seventeen UK CWR hot spots using the iterative method (numbers of CWR taxa present shown).

area network of UK SACs or SSSIs. However, it is unclear if this situation could be replicated in all but a few other European countries, let alone outside of Europe. As such, identifying the location of genetic reserves for multiple target taxa is likely to be more time- and resource-consuming, involving a substantial field-surveying element to the work, but as has been shown by the GEF-funded projects in the Eastern Mediterranean (Zencirci *et al.*, 1998; ICARDA, 2007) it is achievable.

Identification of potential sites to establish genetic reserves constitutes a significant step forward in the implementation of the national CWR conservation strategy for each country, and globally it will ensure the survival of a vital natural resource in a time of growing ecosystem vulnerability. However, identification of potential sites is not an end in itself; once identified, sites need to be supported by targeted long-term governmental resources to ensure that the genetic reserve is fully implemented. Also as noted above, the identification of potential sites always needs to be reviewed in terms of site sustainability prior to establishment of the genetic reserves: factors such as the differential impact on potential sites of climate change are an important consideration as the establishment of a genetic reserve is always long-term and establishing a site that will be lost to rising sea level or desertification would not be sustainable.

2.5 Reserve Design

2.5.1 Optimal reserve design

Although the theory related to the design of reserves has been largely developed by animal conservationists rather than botanists or plant geneticists, the basic principles do not vary a great deal. The current consensus view of an optimal reserve design is the one based on the Man and the Biosphere programme (UNESCO), as discussed by Cox (1993), modified from Batisse (1986). This establishes a central core area with a stable habitat, surrounded by a buffer zone and outside this, where possible, a transition zone (see Fig. 2.4).

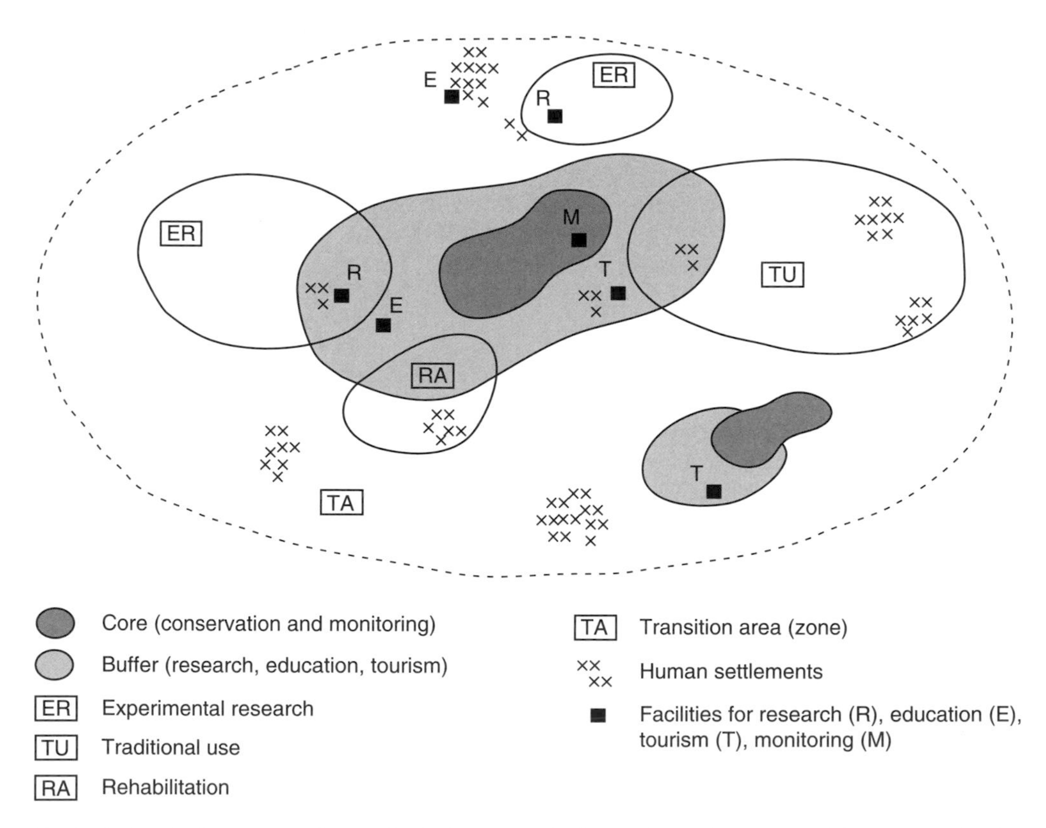

Fig. 2.4. Model for reserve design, including core, buffer and transition zones. The core, typically a protected area such as a national park, is designed to provide for conservation of biotic diversity, with research being limited to monitoring activities. The buffer zone is a region in which research, tourism, educational activities and traditional subsistence activities of indigenous peoples are emphasized. The transition zone, its outer limit not rigidly defined, constitutes an area where major manipulative research, ecosystem restoration and application of research to management of exploited ecosystems are carried out. (Adapted from Batisse, 1986.)

This plan assumes that the core area will be large enough to accumulate an MVP of 1000–5000 potentially breeding individuals already growing as part of the natural ecosystem. This zone will be used for regular monitoring and will be open only to permit-holding scientists and officials. The zone outside this, the buffer zone, protects the core from edge effects and other factors, such as potential conflicts with local communities (Budhathoki, 2004), that might threaten the viability of the target populations in the core. Given (1994) distinguishes two types of buffer zone: extension buffering and socio-buffering. Extension buffering allows effective extension of the core area, thus permitting much larger than theoretically required interbreeding populations to survive. It has a similar management regime to the reserve core and is open to bona fide non-destructive visitors: researchers, educationists and special ecotourists. In contrast, socio-buffering involves the separate management of the core and the buffer, because some sustainable agriculture and forestry is permitted within the buffer zone. It may be advisable to establish a socio-buffer zone if the local people have lost traditional harvesting rights in the reserve core (see Table 2.1). However, it must be stressed that the activities of local people in the buffer zone must be closely monitored to ensure they do not threaten the populations in the reserve core. The benefits of using buffer zones are summarized by Given (1994), modified from Oldfield (1988) and given in Table 2.1.

Outside of the buffer zone, the transition zone will be available for research on ecosystem restoration and similar studies, limited human settlements and sustainable utilization, as well as general tourist visits. This optimal reserve design attempts to involve rather than exclude and alienate local people, while providing a practical buffer and transition zone between the core reserve and areas of intense human activity. It also provides a practical solution to the problem of the boundary between

Table 2.1. Biological and social benefits of using buffer zones. (Adapted from Given, 1994.)

Biological benefits	Social benefits
Provides extra protection from human activities for the strictly protected core zone	Gives local people access to traditionally utilized species without depleting the core area
Protects the core of the reserve from biological change	Compensates people for loss of access to the strictly protected core zone
Provides extra protection from storm damage	Permits local people to participate in conservation of the protected area
Provides a large forest or other habitat unit for conservation, with less species loss through edge effects	Makes more land available for education, recreation and tourism, which in turn helps in conservation advocacy
Extends habitat and thus population size for species requiring more space	Permits conservation of plants and animals to become part of local and regional rural development planning
Allows for a more natural boundary – one relating to movements of animal species that may be essential to some plants	Safeguards traditional land rights and conservation practices of local people
Provides a replenishment zone for core area species including animals essential to some plants	Increases conservation-related employment

areas of conservation reserves and human exploitation. Wherever possible, additional land should be acquired adjacent to the reserve to increase the overall reserve size and to act as additional buffer and transition zones.

The issue of defining a buffer zone and a transition zone may be simplified when the genetic reserve is selected for a plant population or a group of populations that occur within an existing protected area, typically much larger than the core zone of the genetic reserve. In these cases the larger boundaries of the protected area may act as buffer and transition zones for the genetic reserve. A simple modification of the management plan of the protected area may serve for this purpose.

2.5.2 Reserve size

In practice, the size of a reserve is often dictated by the relative human population density and the suitability of the land for human exploitation (agriculture, urbanization, logging, etc.). The less populated or the less productive a country or region, the larger is its reserve size. Primack (1993) provides two examples of large reserves which illustrate this point: the Greenland National Park, which is composed of a frozen land mass of 700,000 km^2; and the Bako National Park in Malaysia, which is set on nutrient-deficient soils. In contrast, lands with economic value for exploitation have few and small reserves. Given (1994) also illustrates this point by listing the 15 largest reserves found in the USA, all in agriculturally marginal areas. Although there may be a correlation between marginal lands and the lands that governments are willing to designate for reserves, there is unlikely to be a natural correlation between marginal agricultural land and the distribution of species worthy of conservation.

However, given the ideal situation where reserve sites are selected on the basis of scientific principles, the ideal size for a reserve has proved to be a continuing point of discussion (Soulé, 1987). Size is commonly related to theories of island biogeography (see MacArthur and Wilson, 1967; Shafer, 1990; Spellerberg, 1991; Cox, 1993) and relative rates of colonization and extinction per unit area. The debate is often centred on the relative advantage of a single large versus several small reserves, the so-called SLOSS debate (Single Large Or Several Small). For example, is it better to have one large reserve of 15,000 ha or a network of five 3000 ha reserves? Large reserves obviously enable a more ecogeographically diverse environment to be included within a single location with minimal edge effect. Alternatively, if a network of smaller reserves is established, each reserve could be sited in a distinct environment, which would better enable the conservation of extreme ecotypes. So the conservation value of multiple small reserves can be greater than the sum of its individual components, especially if the reserves are closely juxtaposed or connected (see below). This debate is often made with regard to maximizing species diversity within a reserve, but similar arguments would also be valid for maximizing genetic diversity of target species for which reserves are being set up. One can also argue that if reserves are too small or too isolated, the populations of the target taxa they contain will become unviable.

As an alternative to the SLOSS debate, Laguna *et al.* (2004) suggested that large and small reserves should be considered as complementary models. In some cases, they may even be compatible systems where, for instance, a large reserve (i.e. a nature park or a national park) for one or a few umbrella species can indirectly provide protection for a large number of 'minor' species. In such a case, the reserve may contain a network of smaller areas for a selected range of plant biodiversity. The equilibrium between complementarity and representativeness can depend on the scenarios planned by the reserve designer and may be influenced by external factors such as feasibility of legal protection, and socio-economical conditions, which often dictate the size of reserves (see the revision and analysis made by Kati *et al.* (2004)).

The current consensus is that the optimal number and size of reserves will depend on the characteristics of the target species and their habitat requirements (Soulé and Simberloff, 1986). Large reserves are better able to maintain species and population diversity because of their greater species and population numbers and internal range of habitats (Abele and Conner, 1979). They allow the physical integrity of the environment to be maintained (e.g. watersheds, drainage system) and are especially suited for low-density species, such as many forest trees, which are commonly found in disjoint populations. However, small, multiple reserves may be more appropriate for annual plant species, which are naturally found in dense but restricted stands (Lesica and Allendorf, 1992). Practically, the vast majority of the wild relatives of crop plants are much more widely distributed. In such cases, a single reserve (or a single cluster of sites) would not suffice. A number of reserves, located in different segments of the distribution area of the target species, would be required to cover its ecogeographic divergence and to deal adequately with the genetic changes which occur over its geographic range.

Within reason, the larger the reserve size, the more diverse are the habitat variants included, which among other things should help ameliorate short-term climatic fluctuations such as drought, heavy rainfall, fungal and insect attack, and other variables (Dinoor *et al.*, 1991; Namkoong, 1991; Nevo *et al.*, 1991; Noy-Meir *et al.*, 1991a,b; Ortega, 1997; Anikster *et al.*, 1998). The reserve should be sufficiently large to permit genetic divergence between populations in space and also in time (Brown, 1991) and to promote gene flow where necessary (Golenberg, 1991). However, there is no point in making the core zone of genetic reserve any larger than the extent of the population or populations.

It should be noted that the same level of deleterious human activity could decimate a small reserve, while a larger reserve may still remain viable. Smaller reserves will generally require more intensive management and monitoring to maintain the same population levels and diversity because of their inherent artificial nature and their limiting capacity to sustain larger population size and availability of suitable habitat. If multiple small reserves are established, each should have sufficient buffering to protect the core from catastrophes. Small reserves, which are often associated with urban areas, can serve an additional purpose, as an educational and nature study centre that will increase public awareness of the importance of conservation.

An application of the model of networks of small reserves for plant conservation is the case of the PMRs network of the Valencian Community (Laguna, 1999, 2001). This network was established in 1991, to capture at least one population of each rare, endemic or threatened wild species in the region of Valencia, Spain, on public sites and/or on private lands whose landowners were keen to ensure long-term management of their sites for nature conservation. The project was developed with the support of the LIFE-Nature programme of the European Commission in 1994 and the first reserves were officially declared by the end of 1997. Currently, the Valencian PMRs network is composed of 258 protected sites, each one extending to 20 ha. They are strictly protected but the maintenance of those traditional activities compatible with the conservation of the target species is permitted. The sites contain about 1030 populations of 450 targeted species which represents more than 85% of the 350 Spanish endemic plant species in the Valencian region. These populations are effectively protected *in situ* and most of them are the object of recurrent monitoring and management practices (Laguna, 2004, 2005). In addition, the Valencian government provides financial support to the landowners of the network to develop their projects of nature conservation such as site restoration, boardwalks and educational exhibits.

Theoretical approaches for single-species reserves (McCarthy *et al.*, 2005) suggest that the optimal reserve configuration depends on whether the management objective is to maximize the mean time to extinction or minimize the risk of extinction. When maximizing the mean time to extinction, the optimal number of independent reserves does not depend on the amount of available habitat for the reserve system. In contrast, the risk of extinction is minimized when individual reserves are equal to the optimal patch size, making the optimal number of reserves linearly proportional to the amount of available habitat.

Some recent studies on economic considerations in reserve design (Groeneveld, 2005) reveal that the socially optimal number of reserve sites (which maximizes social welfare) is generally larger than the ecologically optimal number (which maximizes an ecological objective such as population viability). However, when the opportunity costs of conservation can be offset by land transactions, the socially optimal number of reserve sites might be closer to the ecological optimum. The examples of the Valencian PMR model tend to support the results of Groeneveld's study. Most of the PMRs are public sites, where the Valencian regional government follows a policy of selection of high-quality sites based on scientifically sound criteria. For instance, they would choose sites which are large enough to maintain populations of endemic taxa of the region in the long term. However, in the case of private PMRs, which are currently 41 sites of the PMR network, the selection is made by a public call of proposals, and the long-term engagement of the landowners with plant conservation philosophy becomes an important decision factor. As a result, the private PMRs can have less botanical or ecological quality, but may play a more important role in creating awareness of plant conservation among them and the public. In fact, one of the most valued successes of the PMR network has been the establishment of an association of PMR landowners, called 'Espacios para la Vida'.

2.5.3 Population size

It is more objective and appropriate to target the ideal numbers of individuals that form a viable population, the EPS, that will ensure the effective conservation of genetic diversity in the target species for an indefinite period. Shafer (1990) defines the minimum population size or MVP size for any given habitat as the smallest population having a 99% chance of remaining extant for 100 years. It is difficult to make this estimation with any degree of accuracy as it is compounded by events such as breeding success, predation, competition, disease, genetic drift, interbreeding, founder effect and natural catastrophes, e.g. fire, drought and flooding, which may affect population longevity (Shafer, 1990).

It is generally agreed that a severe reduction in population size will result in loss of alleles with possible negative effects on survival (Shafer, 1990). There are examples known of populations diminishing to a very small size and yet surviving and expanding later, but these 'phoenix' populations will undoubtedly have suffered extensive genetic drift during the process (Lawrence and Marshall, 1997). Precise estimates of the MVP size vary. Frankel and Soulé (1981) mention numbers of individual plants from 500 to 2000. Hawkes (1991) recommends an MVP of at least 1000 individuals, taking into account that not every individual may contribute genes to the next generation. There is an extensive discussion of minimum population size or MVP provided by Lawrence and Marshall (1997), who tentatively conclude that an MVP of 5000 is appropriate in terms of *in situ* genetic conservation. Complementarily, special cases also must be considered, such as species propagating by means of asexual reproduction (bulbs, rhizomes, etc.) or those basing their long-term conservation on the longevity of a few individuals (i.e. some cases of conifers). In any case, if the population is under the most repeated numbers, typically 500 for most authors, the reserve designer is advised to propose complementary measures for *ex situ* conservation, as well as the search for alternative sites to create, if necessary, future safe populations by means of reinforcement or reintroduction techniques.

It is important to bear in mind that a population's minimum viable size depends on many unpredictable factors which influence the persistence of the population at the site. Catastrophes such as continuous severe drought, the appearance of a destructive new pest, hurricanes and fire can reduce the population in a given reserve to just a very small fraction of its original size. In Mediterranean environments, climatic differences between years can be very wide and these factors strongly affect population sizes. The MVP also varies from species to species depending on the mode of reproduction and life form. Annuals seem to be much more vulnerable to such changes compared to perennial species. Two or three years of continuous, severe drought can reduce annual populations to only a small percentage of their average size. Native perennial bushes and trees show much smaller fluctuations. The risks for small annual populations (comprising only a few thousands of individuals) are even greater when this life form is coupled with self-pollination (which is the case in the majority of the wild relatives of Old World grain crops such as wheat, barley, peas or chickpea). In such species, where genetic variation is structured in true breeding lines, severe decimation in small populations will result in loss of genetic diversity, particularly of the less frequent genotypes. As only a limited number of individuals can exist in a small reserve area, the genetic flexibility (the formation of new lines) will also be impaired. For these species larger population and reserve sizes will be required, as is the case for species that are associated with disturbed rather than climax habitats.

In reserve design, a minimal population size should be regarded only as a last resort and it is recommended that a much larger population (up to 10,000 are desirable) should be considered for *in situ* genetic conservation as a safety target. Frequently, this is also a spatial necessity since the area actually needed is usually much larger than that supporting (theoretically) only a few thousand individuals.

2.5.4 Corridors, networks and stepping stones

It is generally accepted that reserve network is a cornerstone of global conservation and resource management (Kati *et al.*, 2004; Meir *et al.*, 2004; Wilson *et al.*, 2005). If multiple small reserves are selected, their potential conservation value can be enhanced by producing a coordinated management plan that attempts to facilitate gene flow and migration between the component reserves. In this way, individual populations can be effectively managed at a metapopulation level. Gene flow may be further advanced by the use of habitat corridors linking individual reserves. These ideas were initially developed for animal conservation, but work equally well for plant species, especially those that rely on animal-based seed dispersal. In terms of animal conservation, the larger the animal, the wider is the corridor required. So if a plant species uses a large animal for seed dispersal, the corridor will need to be of an appropriate width to permit migration of the animal-dispersal agent. In some cases, critical areas may occur where the connectivity by corridors is constrained for certain species, either because of bottlenecks caused by human activities or due to natural landscape and biotic features. Once identified, natural bottlenecks can be managed to retain their inherent connectivity. Human-caused bottlenecks can only be altered again to restore the connectivity they have lost. Habitat corridors, however, do have some conservation drawbacks, because, by definition, they have a high edge to area ratio (see below), they facilitate the rapid distribution of pests and diseases between the reserves, and are rarely feasible in highly fragmented or disturbed habitats.

In an attempt to facilitate interpopulation gene flow in regions of high habitat fragmentation, the GIS technique of Opportunity Mapping has been developed to identify where suitable habitat islands already exist or where degraded habitats might be enhanced to provide stepping stones for gene flow (Saunders and Parfitt, 2005). While corridors are a continuous strip of habitat that differs from the surrounding habitat on either side and effectively provide a strip between patches along which migration and ultimately gene exchange can occur, stepping stones are not contiguous areas of habitat, but are a series of patches close enough to each other to permit the migration of species (Forman and Godron, 1986). The relative effectiveness of both corridors and stepping stones varies according to the species under study (Haddad, 2000). The dynamics of the 'sink' populations (those existing in the corridor or stepping-stone habitat) are affected by the sizes of the source areas (areas of existing habitat), the proximity of these areas to the sink patches and the nature of the surrounding habitat matrix (Wiens, 1989). Certain species are particularly suited to stepping-stone dispersal, such as the rare Fender's Blue butterfly in Oregon in the USA, which shows a preference for hopping between patches of flowers, and therefore habitat stepping stones have proven effective for its conservation (Jensen, 1998). An important consideration is the size and number of patches needed for effective migration and the distances between these patches. The larger each patch is, the fewer will be required, and the greater the number of patches, the more efficient will be the network of stepping stones because of the increased likelihood that a suitable patch will be located and colonized by the migrating species. The effective distance between patches is a more difficult factor to determine, is likely to require field experimentation and is commonly species-specific.

2.5.5 Reserve shape

Reserve shape is often the last of the concerns during reserve design as other factors tend to dictate the final shape a reserve takes. Nevertheless, considerations of the shape of the reserve are important when designing a genetic reserve. Ideally, the edge to area ratio should be kept to a minimum to avoid deleterious micro-environmental effects including changes in light, temperature, wind, incidence of fire, introduction of alien species, grazing, as well as deleterious anthropomorphic effects. The most optimal shape would be a round reserve where the edge to area ratio is kept at a minimum. Long linear reserves will have the highest edge to area ratio and should be avoided unless dealing with islands or territories with clearly defined physical boundaries. Fragmentation of the reserve by roads, fences, pipelines, dams, agriculture, intensive forestry and other human activities will necessarily fragment and limit the effective reserve size, multiply the edge effects and may leave populations in each fragment unsustainable.

2.5.6 Political and economic factors

The design of a reserve must take into consideration the socio-economic factors affecting the chosen site. There are often conflicting uses of the same land for different purposes (e.g. wild plants harvesting, agriculture, infrastructure development and nature conservation). Reserve design should pragmatically be applied to allow complementary use as agricultural, industrial or recreational resources.

Reserve use – No reserve will be successful if the needs of the user groups, such as local population, the general public, reserve visitors and the scientific community, are not considered when designing and managing the reserve. Reserve design should take into account the needs of local communities, local farmers, landowners and other members of the local population who may depend on the proposed reserve site for their livelihood. During the planning phase of the reserve, consultations should also be held with regional and national governments as well as with local communities, thus ensuring that establishment and management agreements are in place before the reserve is functional. Such negotiations should consider the protection status and the legal framework of the reserve. In case the site is not protected, measures to ensure the future protection of the site should be discussed and implemented. In any case, a long-term management or custodianship agreement with the landowners (whether public or private) while considering the needs of local people dependent on the land is essential to ensure the long-term sustainability of the reserve. If the proposed reserve is sited in a region of woodland, desert scrub or bushland which is used by local communities but not actually owned by them, detailed consultations and agreements must also be undertaken with them. These preliminary measures should apply also to hunter-gatherer community areas or regions of shifting cultivations and wild plant 'harvesting'. In all such regions or areas a sympathetic and cooperative attitude should be adopted to ensure that the local communities are fully aware of the importance of the proposed reserve(s). Ideally, reserve staff should be recruited locally and the whole community should be encouraged to take pride in local conservation work. Meffe and Carroll (1994) discuss in detail the rehabilitation of the Guanacaste National Park in Costa Rica and refer to the project based on a philosophical approach called biocultural restoration, which incorporates local people in all aspects of the reserve's development and protection. The project has not only resulted in the restoration of degraded habitats, but has also restored to the local people a biological and intellectual understanding of the environment in which they live and has achieved the goal of creating a 'user-friendly' reserve that contributes to their quality of life.

In some cases, the conservation of a species is fully dependent on the maintenance of local agricultural or farming practices. For instance, the current data of most national plant red data books in Europe show that the moderately nitrophilous plants (including arable plants) represent one of the most sensitive groups of endangered plants, and their wild populations are quickly declining, as nitrogen levels change due to agricultural practices. In most cases it concerns non-protected plants associated to agricultural crops, especially cereal and neglected crops, whose conservation depends on the maintenance of some traditional practices. Such practices are also rapidly disappearing throughout Europe, but are now effectively supported by conservationist NGOs, or by several governmental programmes such as the case of orchid grasslands conservation in White Carpatians, Czech Republic, reviewed by Beckmann (2000).

Reserve design should pay particular attention to cater for the needs of the general population at large, whether local, national or international, who visit the reserve for aesthetic reasons and may support the long-term political and financial viability of the reserve. The growth of public awareness of conservation issues has undoubtedly stimulated a growth in conservation activities and market opportunities for local people in terms of ecotourism in recent years. It is important to keep the public informed of the reserve activities in order to retain public support for the reserve. Reserve design must take into account the needs of visitors, such as, visitors' centres, nature trails and lectures. They are also likely to bring additional income to local people and the reserve through guided tours and the sale of various media reserve information packs.

Special consideration needs to be given in reserve design to scientific documentation and research. One of the assumed values of *in situ* conservation is that the target species will be evolving with general environmental changes and in particular with the constantly changing biotypes of pests and diseases and these evolutionary changes are generally very slow. In this context, an important consideration for reserve design is the future impact of climate change, and corridors and other design aspects to give flexibility to the reserve should be provided as precautionary strategies (Millennium Ecosystem Assessment, 2005).

Reserve sustainability – Establishing and managing an *in situ* genetic reserve is resource-expensive and therefore both reserve and target taxa must be sustainable over an extended period of time to make the investment worthwhile. Reserve sustainability depends on properties of target taxa, MVP size, reserve size and shape as well as management of the reserve. The larger the MVP and reserve size, the higher is the reserve sustainability. Sustainability also has a better perspective when genetic reserves are designed as part of reserve networks that include high-quality sites (Cabeza *et al.*, 2004). The relative cost and ease of establishing reserves will also affect the choice of reserve sites and their design. The choice between two equally appropriate sites would need to be made based on cost–benefit analysis of the two reserves with regard to the conservation budget available, the relative establishment costs, such as cost of land, current usage, meeting the needs of local people, implementing recovery plans and incentive costs. These costs can be very significant and will be the decisive factor. In addition, it may be necessary to consider incentives to be provided to the landowners. These incentives may be direct in terms of cash or in kind, or indirect such as development interventions like community-based natural resource management projects in the site. The Millennium Ecosystem Assessment (2005) pointed out that direct incentives for conservation were more efficient, effective and equitable than indirect payments. The genetic reserve system should be designed in a cost-effective and efficient manner and in most cases common sense will dictate that it will be preferable to conserve multiple target taxa in the same reserve.

References

Abele, L.G. and Conner, E.F. (1979) Application of island biogeography theory to refuge design: making the right decision for the wrong reasons. In: Linn, R.M. (ed.) *Proceedings of the First Conference on Scientific Research in the National Parks.* National Parks Service, Washington, DC.

Anderson, R., Gomez-Laverde, M. and Peterson, A. (2002) Geographical distributions of spiny pocket mice in South America: insights from predictive models. *Global Ecology and Biogeography* 11, 131–141.

Anikster, Y., Horovitz, A. and Noy-Meir, I. (1998) Thirteen years of *in situ* research on wild wheat in Israel. In: Zencirci, N., Kaya, Z., Anikster, Y. and Adams, W.T. (eds) *The Proceedings of International Symposium on* In Situ *Conservation of Plant Genetic Diversity*. Central Research Institute for Field Crops, Ulus, Ankara, Turkey, pp. 55–65.

Araújo, M.B. and Rahbek, C. (2006) How does climate change affect biodiversity? *Science* 313, 1396–1397.

Araújo, M.B., Cabeza, M., Thuiller, W., Hannah, L. and Williams, P.H. (2004) Would climate change drive species out of reserves? An assessment of existing reserve-selection methods. *Global Change Biology* 10, 1618–1626.

Araújo, M.B., Pearson, R.G., Thuiller, W. and Erhard, M. (2005a) Validation of species-climate impact models under climate change. *Global Change Biology* 11, 1504–1513.

Araújo, M.B., Whittaker, R., Ladle, R. and Markus, E. (2005b) Reducing uncertainty in projections of extinction risk from climate change. *Global Ecology Biogeography* 14, 529–538.

Balmford, A. (2002) Selecting sites for conservation. In: Norris, K. and Pain, D. (eds) *Conserving Bird Biodiversity. General Principles and their Application*. Cambridge University Press, Cambridge, pp. 74–104.

Balmford, A., Gaston, K.J., Blyth, S., James, A. and Kapos, V. (2003) Global variation in terrestrial conservation costs, conservation benefits and unmet conservation needs. *Proceedings of the National Academy of Sciences USA* 100, 1046–1050.

Balvanera, P., Daily, G.C., Ehrlich, R.R., Ricketts, T.H. and Bailey, S.A. (2001) Conserving biodiversity and ecosystem services. *Science* 291, 2047.

Batisse, M. (1986) Developing and focussing the biosphere reserve concept. *Nature and Resources* 22, 1–10.

Beckmann, A. (2000) *Caring for the Land. A Decade of Promoting Landscape Stewardship in Central Europe*. NP Agentura, Stare Mesto, Czech Republic.

Bengtsson, J., Angelstam, P., Elmqvist, T., Emanuelsson, U., Folke, C., Ihse, M., Moberg, F. and Nyström, M. (2003) Reserves, resilience and dynamic landscapes. *Ambio* 32(6), 389–396.

Bioversity (2007) Role of geographic analysis in plant genetic diversity. Available at: http://www.bioversityinternational.org/Regions/Americas/pdf/MIE_395_CH17.pdf.

Boffa, J.M. (2000) West African agroforestry parklands: keys to conservation and sustaining management. *Unasylva* 51, 11–17.

Bonn, A., Rodrigues, A.S.L. and Gaston, K.J. (2002) Threatened and endemic species: are they good indicators of patterns of biodiversity on a national scale? *Ecology Letters* 5, 733–741.

Brown, A.H.D. (1991) Population divergence in wild crop relatives. *Israel Journal of Botany* 40, 512.

Brown, A.H.D. and Briggs, J.D. (1991) Sampling strategies for genetic variation in *ex situ* collections of endangered plant species. In: Falk, D.A. and Holsinger, K.E. (eds) *Genetics and Conservation of Rare Plants*. Oxford University Press, New York, pp. 99–119.

Brown, A.H.D. and Schoen, D.J. (1992) Plant population genetic structure and biological conservation. In: Sandlund, O.T., Hindar, K. and Brown, A.H.D. (eds) *Conservation of Biodiversity for Sustainable Development*. Oxford University Press, London.

Brown, A.H.D. and Marshall, D.R. (1995) A basic sampling strategy: theory and practice. In: Guarino, L., Ramanatha Rao, V. and Reid, R. (eds) *Collecting of Plant Genetic Diversity*. CAB International, Wallingford, UK, pp. 75–92.

Budhathoki, P. (2004) Linking communities with conservation in developing countries: buffer zone management initiatives in Nepal. *Oryx* 38(3), 334–341.

Cabeza, M., Araujo, M.B., Wilson, R.J., Thomas, C.D., Cowley, M.J.R. and Moilanen, A. (2004) Combining probabilities of occurrence with spatial reserve design. *Journal of Applied Ecology* 41(2), 252–262.

Center for Plant Conservation (1991) Genetic sampling guidelines for conservation collections of endangered plants. In: Falk, D.A. and Holsinger, K.E. (eds) *Genetics and Conservation of Rare Plants*. Oxford University Press, New York, pp. 225–238.

Cox, G.W. (1993) *Conservation Ecology*. W.C. Brown, Dubugue, Iowa.

Cumming, G. (2000) Using between-model comparisons to fine-tune linear models of species ranges. *Journal of Biogeography* 27, 441–455.

Davis, S.D., Droop, S.J.M., Gregerson, P., Henson, L., Leon, C.J., Lamlein Villa-Lobos, J., Synge, H. and Zantovska, J. (1986) *Plants in Danger: What Do We Know?* International Union for Conservation of Nature and Natural Resources. Gland, Switzerland.

de Klemm, C. (1990) *Wild Plant Conservation and the law*. Environmental Policy and Law Paper, No 24 IUCN, Gland, Swtizerland/Cambridge.

de Klemm, C. (1996) Legal instruments for the protection of wild flora. In: Newton, J. (ed.) *Planta Europa Proceedings*. Plantlife, London, pp. 29–50.

de Klemm, C. (1997) *Comparative Analysis of the Effectiveness of Legislation for the Protection of Wild Flora in Europe*. Nature and Environment series, No. 88, Council of Europe, Strasbourg, France.

de Klemm, C. and Shine, C. (1993) *Biological Diversity Conservation and the Law: Legal Mechanisms for Conserving Species and Ecosystems*. Environmental Policy and Law Paper No 29, IUCN, Gland, Switzerland/Cambridge.

de Vicente, C. (2004) *The Evolving Role of Genebanks in the Fast Developing Field of Molecular Genetics*. Issues in genetic resources, No 11, International Plant Genetic Resources Institute, Rome, Italy.

Dinoor, A., Eshed, N., Ecker, R., Gerechter-Amitai, Z., Sokl, Z., Manisterski, J. and

Anikster, Y. (1991) Fungal diseases of wild tetraploid wheat in a natural strand in northern Israel. *Israel Journal of Botany* 40, 481–500.

Draper, D., Rossello-Graell, A., Garcia, C., Gomes, C. and Sergio, C. (2003) Application of GIS in plant conservation programmes in Portugal. *Biological Conservation* 113, 337–349.

Dulloo, M.E., Verburg, J., Paul, S.S., Green, S.E., de Boucherville-Baissac, P. and Jones, C. (1997) *Ile aux Aigrettes Management Plan 1997–2001*. Technical Series No. 1/97, Mauritian Wildlife Foundation, Mauritius.

Elith, J., Graham, C.H., Anderson, R.P., Dudík, M., Ferrier, S., Guisan, A., Hijmans, R.J., Huettmann, F., Leathwick, R., Lehmann, A., Li, J., Lohmann, L.G., Loiselle, B.A., Manion, G., Moritz, C., Nakamura, M., Nakazawa, Y., Overton, J.McC., Peterson, A.T., Phillips, J., Richardson, K., Scachetti-Pereira, R., Schapire, E., Soberon, J., Williams, S., Wisz, M. and Zimmermann, E. (2006) Novel methods improve prediction of species' distributions from occurrence data. *Ecography* 29, 129–151.

Euro+Med PlantBase (2002) A guide for contributors of initial taxonomic accounts Version 2.0 July 5, 2002. Available at: http://www.euromed.org.uk/d_ocuments/5.7.02_revision_guidelines.pdf.

Ferguson, M.E., Ford-Lloyd, B.V., Robertson, L.D., Maxted, N. and Newbury, H.J. (1998) Mapping the geographical distribution of genetic variation in the genus *Lens* for the enhanced conservation of plant genetic diversity. *Molecular Ecology* 7, 1743–1755.

Ford-Lloyd, B.V., Maxted, N., Kell, S.P., Mitchell, M., Scholten, M. and Magos Brehm, J. (2007) Establishing conservation priorities for crop wild relatives. In: Maxted, N., Ford-Lloyd, B.V., Kell, S.P., Iriondo, J.M., Dulloo, E. and Turok, J. (eds) *Crop Wild Relatives Conservation and Use*. CAB International, Wallingford, UK.

Forman, R.T.T. (1987) Emerging directions in landscape ecology and applications in natural resources management. In: Herrmnn, R. and Craig, T.B. (eds) *Conference on Science in National Parks*. The George White Society and the National Parks Service, Washington, DC, pp. 59–88.

Forman, R.T.T. and Godron, M. (1986) *Landscape Ecology*. Wiley, Chichester, UK.

Frankel, O.H. and Bennett, E. (1970) *Genetic Resources in Plants – Their Exploration and Conservation*. Blackwell, Oxford.

Frankel, O.H. and Soulé, M.E. (1981) *Conservation and Evolution*. Cambridge University Press, Cambridge.

Franklin, J. (1995) Predictive vegetation mapping: geographic modelling of biospatial patterns in relation to environmental gradients. *Progress in Physical Geography* 19, 474–499.

Frodin, D.G. (2001) *Guide to the Standard Floras of the World*, 2nd edn. Cambridge University Press, Cambridge.

Given, D.R. (1994) *Principles and Practice of Plant Conservation*. Chapman & Hall, London.

Golenberg, E.M. (1991) Gene flow and the evolution of multilocus structures in wild wheat. *Israel Journal of Botany* 40, 513.

Graudal, L., Kjaer, E.D. and Canger, S. (1995) A systematic approach to the conservation of genetic resources of trees and shrubs in Denmark. *Forest Ecology and Management* 73, 117–134.

Graudal, L., Kjaer, E.D., Thomsen, A. and Larsen, A.B. (1997) *Planning National Programmes for Conservation of Forest Genetic Resources. Technical Note No. 48*. Danida Forest Seed Centre, Denmark.

Graudal, L., Yanchuk, A. and Kjaer, E. (2004) National planning. In: FAO, FLD, IPGRI (eds) *Forest Genetic Resources Conservation and Management*, Vol. 1: *Overview, Concept, and some Systematic Approaches*. International Plant Genetic Resources Institute, Rome, Italy, pp. 25–36.

Greuter, W., Burdet, H.M. and Long, G. (eds) (1989) *Med-Checklist. A Critical Inventory of Vascular Plants of the Circum-Mediterranean Countries*, Vols. 1,2,3,4. Geneva, Switzerland/Berlin.

Groeneveld, R. (2005) Economic considerations in the optimal size and number of reserve sites. *Ecological Economics*, 52(2), 219–228.

Guisan, A. and Zimmermann, N. (2000) Predictive habitat distribution models in ecology. *Ecological Modelling* 135, 147–186.

Guisan, A., Edwards, T.C. Jr. and Hastie, T. (2002) Generalized linear and generalized additive models in studies of species distribution: setting the scene. *Ecological Modelling* 157, 89–100.

Haddad, N. (2000) Corridor length and patch colonization by a butterfly. *Junonia coenia. Conservation Biology* 14(3), 738–745.

Hanelt, P. and IPK (2001) *Mansfeld's Encyclopaedia of Agricultural and Horticultural Crops*. Springer, Berlin.

Hansson, L., Fahrig, L. and Merriam, G. (1995) *Mosaic Landscapes and Ecological Processes*. Chapman & Hall, London.

Hawkes, J.G. (1991) International workshop on dynamic *in situ* conservation of wild relatives of major cultivated plants: summary of final discussions and recommendations. *Israel Journal of Botany* 40, 529–536.

Hawkes, J.G., Maxted, N. and Ford-Lloyd, B.V. (2000) *The* ex situ *Conservation of Plant Genetic Resources*. Kluwer, Dordrecht, The Netherlands.

Heywood, V. (2003) Conservation and sustainable use of wild species as sources of new ornamentals. In: Düzyaman, E. and Tüzel, Y. (eds) *Proceedings of the International Symposium on Sustainable Use of Plant Biodiversity to Promote New Opportunities for Horticultural Production Development*. Acta Horticulturae 598, pp. 43–53.

Heywood, V.H.H. and Dulloo, M.E. (2005) In situ *Conservation of Wild Plant Species – a Critical Global Review of Good Practices*. Technical bulletin 11, International Plant Genetic Resources Institute, Rome, Italy.

Hijmans, R.J. and Graham, C.H. (2006) The ability of climate envelope models to predict the effect of climate change on species distributions. *Global Change Biology* 12, 1–10.

Hijmans, R.J., Guarino, L., Cruz, M. and Rojas, E. (2001) Computer tools for spatial analysis of plant genetic resources data: 1. DIVA-GIS. *Plant Genetic Resources Newsletter* 127, 15–19.

Hijmans, R.J., Cameron, S.E., Parra, J.L., Jones, P.G. and Jarvis, A. (2005) Very high resolution interpolated climate surfaces for global land areas. *International Journal of Climatology* 25, 1965–1978.

Hirzel, A., Hausser, J., Hessel, D.C. and Perrin, N. (2002) Ecological-niche factor analysis: How to compute habitat-suitability maps without absence data? *Ecology* 83, 2027–2036.

Hopkinson, P., Evans, J. and Gregory, R.D. (2000) National-scale conservation assessments at an appropriate resolution. *Diversity and Distributions* 6, 195–204.

IBPGR (1985) *Ecogeographic Surveying and* In Situ *Conservation of Crop Relatives*. Report of an IBPGR task force meeting held at Washington, DC. International Board for Plant Genetic Resources, Rome, Italy.

ICARDA (2007) Conservation and sustainable use of dryland agrobiodiversity in Jordan, Lebanon, Syria and Palestinian Authority. Available at: www.icarda.org/Gef/Gef.html.

IPCC (2007) Climate change (2007) The physical science basis. Available at: http:// www.ipcc.ch/.

Isager, L., Theilande, I. and Thomson, L. (2004) People's participation and the role of government. In: FAO, FLD, IPGRI (eds) *Forest Genetic Resources Conservation and Management*, Vol. 1: *Overview, Concept, and some Systematic Approaches*. International Plant Genetic Resources Institute, Rome, Italy, pp. 49–71.

IUCN (2001) *IUCN Red List Categories and Criteria : version 3.1*. IUCN Species Survival Commission, IUCN, Gland, Switzerland/Cambridge, ii + 30 pp.

Jarvis, A., Lane, A. and Hijmans, R.H. (in press) Impacts of climate change on crop wild relatives. *Agriculture Ecosystesms and Environment*.

Jensen, M.N. (1998) Butterfly may use flowery stepping stones. *Science News*, Science Service, Washington, DC.

Jones, P.G. and Gladkov, A. (1999) *A Computer Tool for the Distribution of Plants and other Organisms in the Wild. FloraMap*. CIAT, Cali, Colombia.

Jones, P.G., Beebe, S., Tohme, J. and Galwey, N.W. (1997) The use of geographical information systems in biodiversity exploration and conservation. *Biodiversity and Conservation* 6, 947–958.

Jones, P.G., Guarino, L. and Jarvis, A. (2002) Computer tools for spatial analysis of plant genetic resources data: 2. FloraMap. *Plant Genetic Resources Newsletter* 130, 1–6.

Kanashiro, M., Thompson, I.S., Yared, J.A.G., Loveless, M.D., Coventry, P., Martins-da-Silva, R.C.V., Degen, B. and Amaral, W. (2002) Improving conservation values of managed forests: the Dendrogene Project in the Brazilian Amazon. *Unasylva* 53, 209.

Kati, V., Devillers, P., Dufrene, M., Legakis, A., Vokou, D. and Lebrun, P. (2004) Hotspots, complementarity or representativeness? Designing optimal small-scale reserves for biodiversity conservation. *Biological Conservation* 120(4), 471–480.

Kell, S.P., Moore, J., Ford-Lloyd, B.V. and Maxted, N. (2007) Creating a regional catalogue of crop taxa and their wild relatives: a methodology illustrated using the Euro-Mediterranean region. In: Maxted, N., Ford-Lloyd, B.V., Kell, S.P., Iriondo, J.M., Dulloo, E. and Turok, J. (eds) *Crop Wild Relatives Conservation and Use*. CAB International, Wallingford, UK.

Kirkpatrick, J.B. and Harwood, C.E. (1983) Conservation of Tasmanian macrophyte wetland vegetation. *Proceedings of the Royal Society of Tasmania* 117, 5–20.

Kjaer, E., Amaral, W., Yanchuk, A. and Graudal, L. (2004) Strategies for conservation of forest genetic resources. In: FAO, FLD, IPGRI (eds) *Forest Genetic Resources Conservation and Management, Vol. 1: Overview, Concept, and some Systematic Approaches*. International Plant Genetic Resources Institute, Rome, Italy, pp. 5–24.

Laguna, E. (1999) The plant micro-reserves programme in the region of Valencia, Spain. In: Synge, H. and Akeroyd, J. (eds) *Proceedings Planta Europa 1998, Second European Conference on the Conservation of Wild Plants*. The Swedish Threatened Species Unit and Plantlife, Uppsala, Sweden/ London, pp. 181–185.

Laguna, E. (2001) *The Micro-reserves as a Tool for Conservation of Threatened Plants in Europe*. Nature and Environment series no 121. Council of Europe. Strasbourg, France.

Laguna, E. (2004) The plant micro-reserve initiative in the Valencian Community (Spain) and its use to conserve populations of crop wild relatives. *Crop Wild Relative* 2, 10–13.

Laguna, E. (2005) Micro-reserves as a tool for grassland restoration and conservation in the Valencian Community (Spain). In: Struchov, A. and Kuleshova, J. (eds) *Facets of Grassland Restoration*. The Open Country Series, Biodiversity Conservation Center, Moscow, Russia, pp. 105–120.

Laguna, E., Deltoro, V.I., Perez-Botella, J., Perez-Rovira, P., Serra, L., Olivares, A. and Fabregat, C. (2004) The role of small reserves in plant conservation in a region of high diversity in eastern Spain. *Biological Conservation* 119(3), 421–426.

Lawler, J., Whit, D., Nelson, R. and Blaustein, A.R. (2006) Predicting climate-induced range shifts: model differences and model reliability. *Global Change Biology* 12, 1568–1584.

Lawrence, M.J. (2002) A comprehensive collection and regeneration strategy for *ex situ* conservation. *Genetic Resources and Crop Evolution* 4, 199–210.

Lawrence, M.J. and Marshall, D.F. (1997) Plant population genetics. In: Maxted, N., Ford-Lloyd, B.V. and Hawkes, J.G. (eds) *Plant Genetic Conservation. The* In Situ *Approach*. Chapman & Hall, London, pp. 99–113.

Lawrence, M.J., Marshall, D.F. and Davies, P. (1995) Genetics of genetic conservation. I. Sample size when collecting germplasm. *Euphytica* 84, 89–99.

Lesica, P. and Allendorf, F.W. (1992) Are small populations of plants worth preserving? *Conservation Biology* 6, 135–139.

MacArthur, R.H. and Wilson, E.O. (1967) *The Theory of Island Biogeography*. Princeton University Press, Princeton, New Jersey.

Mackensie, N.L., Belbin, L., Margules, C.R. and Kreighery, G.J. (1989) Selecting representative reserve system in remote areas: a case study in the Nullabor Region, Australia. *Biological Conservation* 50, 239–262.

Mallett, J. (1996) The genetics of biological diversity: from varieties to species. In: Gaston, K.J. (ed.) Biodiversity. *A Biology of Numbers and Difference*. Blackwell, Oxford, pp. 13–53.

Margules, C. and Usher, M.B. (1981) Criteria used in assessing wildlife conservation potential: a review. *Biological Conservation* 21, 79–109.

Margules, C.R. and Pressey, R.L. (2000) Systematic conservation planning. *Nature* 405, 243–253.

Marshall, C.R. and Brown, A.H.D. (1975) Optimum sampling strategies in genetic conservation. In: Frankel, O.H. and Hawkes, J.H. (eds) *Crop Genetic Resources for Today and Tomorrow*. Cambridge University Press Cambridge, pp. 3–80.

Maxted, N. and Kell, S. (1998) Ecogeographic techniques and *in situ* conservation: a case study for the legume genus *Vicia* L. in Turkey. In: Zencirci, N., Kaya, Z., Anikster, Y. and Adams, W.T. (eds) *Proceedings of International Symposium on* In Situ *Conservation of Plant Genetic Diversity*. Central Research Institute for Field Crops, Ulus, Ankara, Turkey, pp. 323–344.

Maxted, N., van Slageren, M.W. and Rihan, J.R. (1995) Ecogeographic surveys. In: Guarino, L., Ramanatha Rao, V. and Reid, R. (eds) *Collecting Plant Genetic Diversity. Technical guidelines*. CAB International, Wallingford, UK, IPGRI, Rome, Italy, pp. 255–287.

Maxted, N., Hawkes, J.G., Ford-Lloyd, B.V. and Williams, J.T. (1997a) A practical model for *in situ* genetic conservation. In: Maxted, N., Ford-Lloyd, B.V. and Hawkes, J.G. (eds) *Plant Genetic Conservation: The* In Situ *Approach*. Chapman & Hall, London, pp. 545–592.

Maxted, N., Ford-Lloyd, B.V. and Hawkes, J.G. (1997b) Complementary conservation strategies. In: Maxted, N., Ford-Lloyd, B.V. and Hawkes, J.G. *Plant Genetic Conservation: The* In Situ *Approach*. Chapman & Hall, London, pp. 20–55.

Maxted, N., Hawkes, J.G., Guarino, L. and Sawkins, M. (1997c) Towards the selection of taxa for plant genetic conservation. *Genetic Resources and Crop Evolution* 44, 337–348.

Maxted, N., Scholten, M.A., Codd, R. and Ford-Lloyd, B.V. (2007) Creation and use of a national inventory of crop wild relatives. *Biological Conservation* 140, 142–159.

McAllister, H.H. and Rutherford, A. (1990) *Hedera helix* L. and H. *hibernica* (Kirch.) Bean (Araliaceae) in the British Isles. *Watsonia* 18(1), 7–15.

McCarthy, M.A., Thompson, C.J. and Possingham, H.P. (2005) Theory for designing nature reserves for single species. *American Naturalist* 165(2), 250–257.

Meffe, G.K. and Carroll, C.R. (1994) *Principles of Conservation Biology*. Sinauer Associates, Sunderland, Massachusetts.

Meir, E., Andelman, S. and Possingham, H.P. (2004) Does conservation planning matter in a dynamic and uncertain world? *Ecology Letters* 7(8), 615–622.

Menendez, R., Gonzalez, A., Hill, J.K., Braschler, B., Willis, S., Collinghan, Y., Fox, R., Roy, D. and Thomas, C.D. (2006) Species richness changes lag behind climate change. *Proceedings Biological Sciences* 273(1593), 1465–1470.

Midgley, G.F., Thuiller, W. and Higgins, S.I. (2007) Plant species migration as a key uncertainty in predicting future impacts of climate change on ecosystems: progress and challenges. In: Canadell, J.G., Pataki, D.E. and Pitelka, L.F. (eds) *Terrestrial Ecosystems in a Changing World*. Springer, Berlin/Heidelberg, pp. 129–137.

Millennium Ecosystem Assessment (2005) *Ecosystem and Human Well-being. Biodiversity Synthesis*. World Resources Institute, Washington, DC.

Naidoo, R. and Ricketts, T.H. (2006) Mapping the economic costs and benefits of conservation. *PLoS Biology* 4(11), 360.

Namkoong, G. (1991) Dynamics of *in situ* conservation: can fragmentation be useful? *Israel Journal of Botany* 40, 518.

Navarro, A., Ferrando, I. and Laguna, E. (2006) Censo y riesgo de extinción del endemismo vegetal valenciano *Limonium dufourii* (Girard) Kuntze. *Toll Negre* 8, 21–26.

Neel, M.C. and Cummings, M.P. (2003a) Genetic consequences of ecological reserve design guidelines: an empirical investigation. *Conservation Genetics* 4, 427–439.

Neel, M.C. and Cummings, M.P. (2003b) Effectiveness of conservation targets in capturing genetic diversity. *Conservation Biology* 17, 219–229.

Neilson, R.P., Pitelka, L.F., Solomon, A.M., Nathan, R., Midgley, G.F., Fragoso, J.M.V.,

Lischke, H. and Thompson, K. (2005) Forecasting regional to global plant migration in response to climate change. *BioScience* 55, 749–759.

Nevo, E. (1988) Genetic resources of wild emmer wheat revisited; genetic evolution, conservation and utilization. In: Miller, T.E. and Koebner, R.M.D. (eds) *Proceedings of the 7th International Wheat Genetics Symposium*. Cambridge, UK, pp. 121–126.

Nevo, E., Noy-Meir, I., Beiles, A., Krugman, T. and Agani, M. (1991) Natural selection of allozyme polymorphisms: microgeographical spatial and temporal ecological differentiations in wild eminar wheat. *Israel Journal of Botany* 40, 419–449.

Newbury, H.J. and Ford-Lloyd, B.V. (1997) Estimation of genetic diversity. In: Maxted, N., Ford-Lloyd, B.V. and Hawkes, J.G. (eds) *Plant Genetic Conservation: The* In Situ *Approach*. Chapman & Hall, London/New York, pp. 192–206.

Niinemets, U., Valladares, F. and Ceulemans, R. (2003) Leaf-level phenotypic variability and plasticity of invasive *Rhododendron ponticum* and non-invasive *Ilex aquifolium* co-occurring at two contrasting European sites. *Plant Cell and Environment* 26, 941–956.

Noss, R.F. and Harris, L.D. (1986) Nodes, networks and mums: preserving diversity at all scales. *Environment Management* 10, 299–309.

Noy-Meir, I., Agami, M., Cohen, E. and Anikster, Y. (1991a) Floristic and ecological differentiation of habitats within a wild wheat population at Ammiad. *Israel Journal of Botany* 40, 363–384.

Noy-Meir, I., Agami, M. and Anikster, Y. (1991b) Changes in the population density of wild eminar wheat (*Triticum turgidum var. dicoccoides*) in a Mediterranean grassland. *Israel Journal of Botany* 40, 385–395.

Odum, E.P. (1971) *Fundamentals of Ecology*. W.B. Saunders, Philadelphia, Pennsylvania/London, pp. 574.

Oldfield, S. (1988) *Buffer Zone Management in Tropical Moist Forests: Case Study and Guidelines*. IUCN, Gland, Switzerland.

Ortega, R. (1997) Peruvian *in situ* conservation of Andean crops. In: Maxted, N., Ford-Lloyd, B.V. and Hawkes, J.G. (eds) *Plant Genetic Conservation. The* In Situ *Approach*. Chapman & Hall, London, pp. 302–314.

Ortega-Huerta, M.A. and Peterson, A.T. (2004) Modelling spatial patterns of biodiversity for conservation prioritisation in North-eastern Mexico. *Diversity and Distributions* 10, 39–54.

Osborn, T.J. and Briffa, K.R. (2005) The spatial extent of 20th-century warmth in the context of the past 1200 years. *Science* 311, 841–844.

Osborne, P. and Suárez-Seoane, S. (2002) Should data be partitioned spatially before building large-scale distribution models? *Ecological Modelling* 157, 249–259.

Parmesan, C. and Yohe, G. (2003) A globally coherent fingerprint of climate change impacts across natural systems. *Nature* 421, 37–42.

Pearce, J. and Ferrier, S. (2000) Evaluating the predictive performance of habitat models developed using logistic regression. *Ecological Modelling* 133, 225–245.

Pearson, R.G. (2006) Climate change and the migration capacity of species. *Trends in Ecology and Evolution* 21(3), 111–113.

Peperkorn, R., Werner, C. and Beyschlag, W. (2005) Phenotypic plasticity of an invasive acacia versus two native Mediterranean species. *Functional Plant Biology* 32, 12.

Phillips, S.J., Anderson, R. and Schapire, R.E. (2006) Maximum entropy modeling of species geographic distributions. *Ecological Modelling* 190, 231–259.

Pickett, S.T.A. and Thompson, J.N. (1978) Patch dynamics and nature reserves. *Biological Conservation* 13, 27–37.

Pressey, R.L. and Nicholls, A.O. (1989) Efficiency in conservation evaluation: scoring versus iterative approaches. *Biological Conservation* 50, 199–218.

Pressey, R.L., Humphries, C.R., Vane-Wright, R.I. and Williams, P.H. (1993) Beyond opportunism: key principles for systematic reserve selection. *Trends in Ecology and Evolution* 8, 124–128.

Pressey, R.L., Possingham, H.P. and Day, J.R. (1997) Effectiveness of alternative heuristic algorithms for identifying indicative minimum requirements for conservation reserves. *Biological Conservation* 80, 207–219.

Primack, R.B. (1993) *Essentials of Conservation Biology*. Sinauer Associates, Sunderland, Massachusetts.

Purseglove, J.W. (1968) *Tropical Crops Dicotyledons*. Longman, London, pp. 719.

Purseglove, J.W. (1972) *Tropical Crops Monocotyledons*. Longman, London, pp. 670.

Pysek, P., Jarosik, V. and Kucera, T. (2002) Patterns of invasions on temperate nature reserves. *Biological Conservation* 104, 13–24.

Rebelo, A.G. (1994) Iterative selection procedures: centres of endemism and optimal placement of reserves. In: Huntley, B.J. (ed.) *Botanical Diversity in Southern Africa*. National Botanical Institute, Pretoria, South Africa, pp. 231–257.

Robertson, M., Caithness, N. and Villet, M. (2001) A PCA-based modelling technique for predicting environmental suitability for organisms from presence records. *Diversity and Distribution* 7, 15–27.

Ruegg, K.C., Hijmans, R.J. and Moritz, C. (2006) Climate change and the origin of migratory pathways in the Swainson's thrush, *Catharus ustulatus*. *Journal of Biogeography* 33, 1172–1182.

Rutherford, A., Mc Allister, H.A. and Mill, R.R. (1993) New Ivies from the Mediterranean area and Macaronesia. *The Plantsman* 15(2), 115–128.

Sala, O.E., Chapin, F.S., Armesto, J.J., Berlow, E., Bloomfield, J., Dirzo, R., Huber-Sanwald, E., Huenneke, L.F., Jackson, R.B., Kinzig, A., Leemans, R., Lodge, D.M., Mooney, H.A., Oesterheld, M., Poff, N.L., Sykes, M.T., Walker, B.H. and Wall, D.H. (2000) Global biodiversity scenarios of the year 2100. *Science* 287, 1770–1774.

Saunders, G. and Parfitt, A. (2005) *Opportunity Maps for Landscape-scale Conservation of Biodiversity: A Good Practice Study*. English Nature Research Reports, Number 641. English Nature, Peterborough, UK.

Schultze-Motel, J. (1966) Verzeichnis forstlich kultivierter Pflanzenarten [Enumeration of cultivated forest plant species]. *Kulturpflanze Beiheft* 4, 486.

Shafer, C.L. (1990) *Nature Reserves, Island Theory and Conservation Practice*. Smithsonian Press, Washington, DC/London.

Shao, G. and Halpin, P.N. (1995) Climatic controls of eastern North American coastal tree and shrub distributions. *Journal of Biogeography* 22, 1083–1089.

Shine, C. (1996) Private or voluntary systems of natural habitats' protection and management. *Nature and Environment Series* No. 85. Council of Europe, Strasbourg, France.

Simberloff, D. (1986) Design of nature reserves. In: Usher, M.B. (ed.), *Wildlife Conservation Evaluation*. Chapman & Hall, London, pp. 316–337.

Simpson, C. and Starr, J. (2001) Registration of 'COAN' peanut. *Crop Science* 41(3), 918.

Soulé, M.E. (1987) *Viable Populations for Conservation*. Cambridge University Press, Cambridge.

Soulé, M.E. and Simberloff, D. (1986) What do genetics and ecology tell us about the design of nature reserves? *Conservation Biology* 35, 19–40.

Spellerberg, I.F. (1991) Biogeographical basis for conservation. In: Spellerberg, I.F., Goldsmith, F.B. and Morris, M.G. (eds) *The Scientific Management of Temperate Communities for Conservation*. The British Ecological Society by Blackwell, Oxford, pp. 293–322.

Stace, C. (1997) *New Flora of the British Isles*. Cambridge University Press, Cambridge.

STAP-GEF (1999) Report of the STAP brainstorming on the use of taxonomic information: Key outcomes and suggestions. Scientific and Technical Advisory Panel (STAP) of the GEF (Global Environmental Facility). STAP Secretariat, UNEP, Paris. Available at: http://www.gefweb.org/COUNCIL/GEF_C14/gef_c14_inf13.doc.

Tan, A. and Tan, A.S. (2002) *In situ* conservation of wild species related to crop plants: the case of Turkey. In: Engels, J.M.M., Ramantha Rao, V., Brown, A.H.D and Jackson, M.T. (eds) *Managing Plant Genetic Diversity*. CAB International, Wallingford, UK; International Plant Genetic Resources Institute (IPGRI), Rome, Italy, pp. 195–204.

Theilade, I., Graudal, L. and Kjær, E. (2000) *Conservation of the Genetic Resources of* Pinus merkusii *in Thailand*. DFSC Technical Note No. 58. Royal Forest Department (RFD), Thailand/Forest Genetic Resources and Management Project (FORGENMAP), Thailand/FAO of the United Nations/Danida Forest Centre, Humlebaek, Denmark.

Theilade, I., Sekeli, P.M., Hald, S. and Graudal, L. (eds) (2001) *Conservation Plan for Genetic Resources of Zambezi Teak* (Baikiaea plurijuga) *in Zambia*. DFSC Case Study No. 2. Danida Forest Seed Centre, Humlebaek, Denmark.

Thomas, C.D., Cameron, A., Green, R.E., Bakkenes, M., Beaumont, L.J., Collingham, Y.C., Erasmus, B.F.N., Ferreira De Siqeira, M., Grainger, A., Hannah, L., Hughes, L., Huntley, B., Van Jaarsveld, A.S., Midgley, G.F., Miles, L., Ortega-Huertas, M.A., Peterson, A.T., Phillips, O.L. and Williams, S.E. (2004) Extinction risk from climate change. *Nature* 427, 145–148.

Thomson, L., Graudal, L. and Kjaer, E. (2001) Selection and management of *in situ* gene conservation areas for target species. In: FAO, DFSC, IPGRI (eds) *Forest Genetic Resources Conservation and Management, in Managed Natural Forest and Protected Areas*, Vol. 2. International Plant Genetic Resources Institute. Rome, Italy.

Thuiller, W., Araújo, M.B., Pearson, R.G., Whittaker, R.J., Brotons, L. Lavorel, S. (2004) Biodiversity conservation: uncertainty in predictions of extinction risk. *Nature* 430, 34.

Thuiller, W., Lavorel, S., Araújo, M.B., Sykes, M.T. and Colin Prentice, I. (2005) Climate change threats to plant diversity in Europe. *Proceedings of the National Academy of Sciences USA* 102(23), 8245–8250.

Turner, W., Spector, S., Gardiner, N., Fladeland, M. and Sterling, E. (2003) Remote sensing for biodiversty science and conservation. *Trends in Ecological Evolution* 18, 306–314.

Tutin, T.G., Heywood, V.H. *et al.* (1964–1988) *Flora Europaea*, Vols. 1–5, Cambridge University Press, Cambridge.

Tutin, T.G. *et al.* (eds) (1993) *Flora Europaea*, Vol. 1, 2nd edn. Cambridge University Press, Cambridge.

Urban, D.L., O'Neill, V. and Shugart, H.H. (1987) Landscape ecology. *BioScience* 37, 119–127.

Valladares, F., Sanchez-Gomex, D. and Zavala, M.A. (2006) Quantitative estimation of phenotypic plasticity: bridging the gap between the evolutionary concept and its ecological applications. *Journal of Ecology* 94, 1103–1116.

Van Jaarsveld, A.S., Midgley, G.F., Scholes, R.J. and Reyers, B. (2003) *Conservation Management in a Changing World*. AIACC Working Paper No. 1. Available at: www.aiaccproject.org.

Vargas, P., McAllister, H.A., Morton, C., Jury, S.J. and Wilkinson, M.J. (1999) Polyploid speciation in *Hedera* (Araliaceae): phylogenetic and biogeographic insights based on chromosome counts and ITS sequences. *Plant Systematics and Evolution* 219, 165–179.

Walker, P. and Cocks, K. (1991) Habitat: a procedure for modelling a disjoint environmental envelope for a plant or animal species. *Global Ecology and Biogeography Letters* 1, 108–118.

Wiens, J.A. (1989) *Processes and Variations. The Ecology of Bird Communities*, Vol. 2. Cambridge University Press, Cambridge.

Williams, D.G., Mack, R.N. and Black, R.A. (1995) Ecophysiology of introduced *Pennisetum setaceum* on Hawaii: the role of phenotypic plasticity. *Ecology* 76, 1569–1580.

Williams, J.T. (1997) Technical and political factors constraining reserve placements. In: Maxted, N., Ford-Lloyd, B.V. and Hawkes, J.G. (eds) *Plant Genetic Conservation: The* In Situ *Approach*. Chapman & Hall, London/New York, pp. 88–98.

Williams, P., Gibbons, D., Margules, C., Rebelo, A., Humphries, C. and Pressey, R. (1996) A comparison of richness hotspots, rarity hotspots and complementary areas for conserving diversity of British birds. *Conservation Biology* 10, 155–174.

Wilson, K.A., Westphal, M.I., Possingham, H.P. and Elith, J. (2005) Sensitivity of conservation planning to different approaches to using predicted species distribution data. *Biological Conservation* 122(1), 99–112.

Wyse Jackson, P.S. and Akeroyd, J.R. (1994) *Guidelines to be followed in the Design of Plant Conservation or Recovery Plans*. Nature and Environment Series No 68. Council of Europe, Strasbourg, France.

Yahner, R.H. (1988) Changes in wildlife communities near edges. *Conservation Biology* 2, 333–339.

Young, A.G. and Clarke, G.M. (2000) *Genetics, Demography and Viability of Fragmented Populations*. Cambridge University Press, Cambridge, 456p.

Zencirci, N., Kaya, Z., Anikster, Y. and Adams, W.T. (1998) *The Proceedings of the International Symposium on* In Situ *Conservation of Plant Genetic Diversity*. Central Research Institute for Field Crops, Ulus, Ankara, Turkey.

3 Genetic Reserve Management

N. Maxted,[1] J.M. Iriondo,[2] L. De Hond,[2] E. Dulloo,[3] F. Lefèvre,[4] A. Asdal,[5] S.P. Kell[1] and L. Guarino[3]

[1]*School of Biosciences, University of Birmingham, Edgbaston, Birmingham, UK;* [2]*Área de Biodiversidad y Conservación, Depto. Biología y Geología, ESCET, Universidad Rey Juan Carlos, Madrid, Spain;* [3]*Global Crop Diversity Trust, c/o FAO, Rome, Italy;* [4]*INRA, URFM, Unité de Recherches Forestières Méditerranéennes (UR629) Domaine Saint Paul, Site Agroparc, Avignon Cedex 9, France;* [5]*Norwegian Genetic Resources Centre, Aas, Norway*

3.1 Introduction

In large protected areas the effects of storms, wildfires, natural cycles of growth, maturation, decay and regeneration, and even sustainable human exploitation would be buffered and would form part of the natural ecosystem dynamic. As such, these areas would have no need for specific conservation intervention or management, as the majority of species would be in balance, extinction equalling regeneration or immigration. However, such large protected areas are rare and primarily restricted to the few remaining truly natural wilderness areas around the globe. Such protected areas are also not the normal domain of crop wild relative (CWR) species, as many agricultural CWR species are more often associated with disturbed, pre-climax communities at earlier stages of succession (Maxted *et al.*, 1997a). Therefore, the reserve in which they are found is likely to be smaller, commonly abutting urban areas and will require active management intervention to maintain the habitat characteristics and prevent the site reaching its natural climax state. The types of management interventions required to maintain the desired habitat and so sustain target CWR species might include vegetation cutting, grazing, burning or other forms of human-mediated disturbance.

As introduced in Chapter 1, plant species conserved *in situ* require 'active' management (Maxted *et al.*, 1997b). Many CWR and other plant species are undoubtedly conserved in numerous existing formally designated protected and non-protected areas worldwide, such as wastelands, field margins, primary forests and national parks, but in each of these cases the existence of a particular species is likely to be coincidental because the site is managed for agriculture, recreation or habitat diversity, or is not managed in any form. In terms of conservation this may be termed 'passive' conservation, where 'healthy' plant populations occur coincidentally within a location without active species management by conservationists. The term 'active' management implies some form of dynamic intervention at the site, even if that intervention were simply limited to an agreement to monitor particular plant populations. Provided there is no deleterious change in the population, no further management intervention would be required, but in the case of passive conservation there is by definition no management or monitoring of the target taxa. As such, these passively conserved populations are not actively monitored, and they are inherently more vulnerable to extinction, i.e. any deleterious environmental trend that would impact a species may go unnoticed and no actions would be adopted to halt population decline. To ensure the target population's health, positive actions may be needed to promote the sustainability of the target taxa and maintain their natural or artificial (e.g. agricultural) ecosystems within the location chosen for the reserve.

The *in situ* conservation of plant genetic diversity implies a high level of target population scrutiny, and the conservationist needs to be assured not only of relative demographic stability but also that, although the target population will continue to evolve, the magnitude of genetic diversity is not dramatically curtailed. So although the preference is often to locate genetic reserves in existing protected areas, those protected areas must be actively managed. If a site is passively managed, as many protected areas are, then the additional management required to maintain plant genetic diversity is unlikely to be forthcoming unless additional resources accompany the designation as a genetic reserve. Therefore, establishing genetic reserves in actively managed existing protected areas, whether in fact a formal designated protected area or more informal roadside or orchard, is likely to be the norm – active management is the critical factor. This implies the need for associated target species and habitat monitoring, management and protection, and in turn these are often dependent on the existence of a management plan that constitutes the major tool to assist the conservationist in ensuring a viable target population that its inherent genetic diversity is maintained.

3.2 Genetic Reserve Management Plans

The writing and implementation of a management plan for a genetic reserve is an essential step in efficient and effective *in situ* genetic conservation of CWR diversity, just as it is for any wild species conserved in a protected area. The primary aim of genetic reserve management plan is to ensure the maintenance or enhancement of the genetic diversity of the target CWR taxa within the reserve. Along with this primary goal, the management plan should also assist in ensuring that the subordinate management goals for the reserve site are met:

- Maintaining maximum genetic diversity of the target taxon and key associated species;
- Promoting general biodiversity conservation and minimizing threat to all levels of diversity;
- Maintaining natural ecological and evolutionary processes that are not deleterious to the target taxon gene pool;
- Ensuring that appropriate, but minimally intrusive, management interventions enhance target taxon diversity;
- Promoting public awareness of the need for genetic and protected area conservation;
- Facilitating the linkage of conservation to sustainable usage by ensuring that diversity is made available for actual or potential utilization.

However, given these objectives it should be recognized that not all species are in fact suitable for conservation in a genetic reserve: (i) species with disparate populations that generally occur at very low local density, such as many tree species, would not form a viable population at any spatial scale compatible with a genetic reserve establishment; (ii) strict pioneer species forming metapopulations where the parents and offspring do not necessarily share the same location, e.g. wild lentil (*Lens culinaris* subsp. *orientalis* (Boiss.) Ponert) in Turkey (Karagöz, 1998) or black poplar (*Populus nigra* L.) along river banks in Europe (Lefèvre *et al.*, 2001); and (iii) highly threatened species with suboptimal population numbers that would not be viable *in situ*. In the first two examples, traditional large-scale nature reserve areas and ecosystem conservation may provide a better solution because of the size of the population required, while in the last example, *ex situ* or *in situ* restoration, and ideally both, provide the most appropriate solution.

For the majority of species that can be conserved in a genetic reserve before establishing the genetic reserve and writing the management plan, those mandated to carry out the conservation must previously have selected the target taxa (see Maxted *et al.*, 1997b), and developed a clear strategic plan, which states the objectives of the conservation activities. As a prerequisite or as part of this, there will be a need to review the target gene pool and to undertake an ecogeographic survey or preliminary survey mission (see Maxted *et al.*, 1995). The ecogeographic survey should conclude with a clear, concise statement of the proposed conservation objectives and priorities, and should identify appropriate strategies and methods for their implementation. Although the ecogeographic survey will have highlighted broad areas of target taxon distribution and diversity, it is unlikely to pinpoint where the actual reserve should be located. This will involve field surveying to identify locations with healthy target taxon populations and then pragmatic decisions on which of the competing sites offers the best chance of sustainability for the genetic reserve.

As well as assisting in the selection of the most appropriate location for establishing the genetic reserve (see Dulloo *et al.*, Chapter 2, this volume), the ecogeographic survey will also provide the general biological information required for the management plan. The ecogeographic survey will help clarify the conservation objective of the site, facilitate the various elements of the taxon description, provide details of the autecology and local community synecology of the target taxon, and support the design of the reserve and suggest how the target taxon population might be utilized once it is conserved. Thus, there are many necessary steps in the conservation process prior to the writing of the management plan.

It should also be noted that there is a fundamental difference between ecosystem-based protected area management and genetic reserve management that is reflected in the style and content of their respective management plans. The difference is associated with the level of biodiversity being addressed: for ecosystem-based protected area management the goal is commonly broader in terms of taxa (entire communities, ecosystems or vegetation types) and monitoring is focused on species' presence/absence or indicator population characteristics (density, frequency and cover) with a goal of assessing overall biotic 'health' of the site, while for genetic reserve management the focus is narrower in terms of taxa (relatively small number of target taxa) and monitoring is based on estimating genetic diversity, as well as species' presence/absence or certain species' population characteristics. As such, the taxonomic focus of the two approaches will also be distinct: for ecosystem-based protected area management the taxon focus is likely to be either keystone or threatened wild species, while for genetic reserve management the taxon focus is likely to be priority CWR taxa that have a socio-economic value associated with their actual or potential use as gene donors. Maintaining genetic diversity is key to the IUCN definition of a viable population as one which: (i) maintains its genetic diversity; (ii) maintains its potential for evolutionary adaptation; and (iii) is at minimal risk of extinction from demographic fluctuations, environmental variations and potential catastrophe, including overuse (IUCN, 1993). Having made these distinctions, with the obvious exception of the genetic diversity elements of the plan, the actual format of the management plan will have obvious similarities.

3.3 Why Is a Management Plan Required?

The reserve site would have been selected because it contained abundant and hopefully genetically diverse populations of the target taxon; if this is the case, why is there a need to spend time writing and updating a management plan? The general answer is because the CWR target taxon population is likely to require active management to maintain diversity, particularly as the majority of CWR taxa are found in pre-climax communities, and the only means of effectively organizing this management is via a carefully constructed management plan.

For CWR species it could be argued that the need for a management plan is more critical than for traditional ecosystem-based conservation because potentially with maintenance of genetic diversity as a goal it would be possible for a CWR population to maintain normal population characteristics (density, frequency and cover) while losing genetic diversity. As such, the management is likely to be more 'active' than for ecosystem-based protected area conservation management where intra-taxon diversity is not the focus of the conservation effort. It is likely that the more active and complex the management, the greater will be the need for a conservation blueprint to form the basis of management interventions; the genetic conservation management plan aims to detail conservation actions and management interventions.

A management plan is also likely to be required because, as noted in Chapter 2, the location of genetic reserves will often result from a compromise between biological best practice and sociopolitical-ethnographic expediency. This compromise in locating genetic reserves means that the management plan can act as a useful tool to balance competing priorities. For example, the reserve area may be too small or fragmented and isolated to support the ideal minimum viable population or permit natural immigration to balance local extinctions. Therefore, intervention management is necessary to increase or maintain populations at viable levels or to translocate individuals between management areas. All of this can be discussed and recommendations made within the management plan.

The reserve may also be surrounded by hostile anthropogenic environments that result in regular introduction of invasive species (weeds, diseases and generalist predators) and degrading processes (siltation and pollution). Therefore, again, intervention management outlined in the plan is necessary to minimize or remove such negative influences. A reserve established in or near an urban environment may be under pressure for development, for release of their natural resources for human use, or in rural areas for use as agricultural lands to feed rapidly increasing and desperately poor human populations, in which case it is only via the application of the management plan that competing biological and sociopolitical-ethnographic factors can be objectively evaluated.

Although many medicinal CWR and forest species are associated with the higher stages of succession, agricultural CWR species are often associated with disturbed, pre-climax communities (Maxted *et al.*, 1997a). Succession is the universal, natural process of directional change in vegetation during ecological time, that runs from bare soil through intermediate vegetation types culminating in a climax community when the ecosystem achieves directional stability (Krebs, 2001). As such, the target taxon may have evolved and developed a 'healthy' population at the reserve location with regular disturbance and at an intermediate stage of succession. In these cases, removal of the disturbance (grazing, fire, mowing, etc.) or allowing succession to the climax community is likely to be to the detriment of the target taxon population. The dynamics of the target taxon population and the ecological relationships with other taxa within the reserve need to be understood and the appropriate active management interventions must be written into the management plan. Lack of active management may result in lack of the necessary disturbance or succession, which, in turn, is likely to harm population viability and diversity.

3.4 Role of Management Plan Following Reserve Establishment

Once established, the management plan will serve multiple purposes in aiding the conservation of the target taxon population. It will describe the physical, sociopolitical-ethnographic and biological environment of the reserve, and so prove to be a reference work for site management. It will articulate the general conservation objectives and specific goals of the individual reserve and how that reserve sits within institutional, national and regional conservation strategy, thus ensuring consistency of implementation. Through its analysis and statement of sociopolitical-ethnographic and biological environment it will facilitate the anticipation of any natural or anthropogenic conflict or problems associated with managing the reserve. It will describe the management objectives and, in as much detail as is possible, the management interventions required for achieving these objectives, as well as the monitoring practices to be implemented. It will assist in organizing human and financial resources, and act as a training guide for new reserve staff. It will facilitate communication and collaboration between the individual reserve site and other genetic reserves, protected areas and *ex situ* conservation facilities. It will also act as guidelines for the use of the target taxon and its genetic diversity, and assist in raising public awareness of the importance of the specific CWR taxon and CWR taxa as a whole.

3.5 Elements of a Genetic Reserve Management Plan

The actual content or style of a genetic reserve management plan will vary depending on the location, target species, organization, staff, etc. that are involved. There is no standard format, but issues commonly addressed are: conservation context and objectives; site abiotic, biotic and anthropogenic description; taxon description; and necessary research agenda and management prescription. Detailed guidelines for preparing ecological management plans are provided by Thomas and Middleton (2003). However, possible elements of a genetic reserve management plan were summarized by Maxted *et al.* (1997c) and these were discussed and amended during the PGR Forum Population Management Methodologies workshop in Minorca, Spain (De Hond and Iriondo, 2004). The proposed elements of a genetic reserve management plan are summarized in Box 3.1.

As the specific focus of establishing the genetic reserve will be to conserve diversity at the taxonomic and genetic levels for the specific target taxon or taxa, the management plan will require details associated with each of these biodiversity levels. As such, there is a requirement to clarify the recognizable taxonomic elements of the target taxa present in the reserve; the species, subspecies and varieties present; and to describe their characteristics (e.g. taxonomy, phenology, habitat preference, breeding system and minimum population size). For these taxa it will be necessary to describe their demographic structure (e.g. mapping of populations size/age structure and density within the site) and autecology at the reserve site.

There is also a requirement to describe the genetic diversity within the target taxa at the site. Just as diverse sites are selected because of the perceived and most often real link between ecogeographic diversity and genetic diversity, so is it likely that within the site, especially a large or ecogeographically diverse site, there is genetic diversity associated with diverse habitats. Thus, when establishing the reserve there will be a need to sample throughout the reserve to identify genetic neighbourhoods, i.e. pockets of unique or varied genetic diversity. These will need to be managed accordingly to promote the maintenance of that diversity and the pockets of unique or varied genetic diversity are likely to form sub-sites for routine genetic monitoring as part of the overall monitoring regime.

Box 3.1. Genetic reserve management plan content.

1. *Preamble*: Conservation objectives, site ownership and management responsibility, reasons for location of reserve, evaluation of populations of the target taxon, reserve sustainability, factors influencing management (legal constraints of tenure and access)

2. *Conservation context*: Place reserve within broader national conservation strategy for the responsible conservation agency and target taxon, likely interaction between target taxon and climate change at site, externalities (e.g. political considerations), obligations to local people (e.g. allowing sustainable harvesting), present conservation activities (*ex situ* and *in situ*), general threat of genetic erosion

3. *Site abiotic description*: Location (latitude, longitude, altitude), map coverage, photographs (including aerial), detailed physical description (geology, geomorphology, climate and predicted climate change, hydrology, soils)

4. *Site biotic description*: General biotic description of the vegetation, flora and fauna of the site, focusing on the species that directly interact with the target taxa (keystone species, pollinators, seed dispersers, herbivores, symbionts, predators, diseases, etc.)

5. *Site anthropogenic description*: Effects of local human population (both within reserve and around it), land use and land tenure (and history of both), cultural significance, public interest (including educational and recreational potential), bibliography and register of scientific research

6. *General taxon description*: Taxonomy (classification, delimitation, description, iconography, identification aids), wider distribution, habitat preferences, phenology, breeding system, means of reproduction (sexual or vegetative) and regeneration ecology, genotypic and phenotypic variation, local name(s) and uses

7. *Site-specific taxon description*: Taxa included, distribution, abundance, demography, habitat preference, minimum viable population size, and genetic structure and diversity of the target taxon within the site, autecology within the reserve with associated fauna and flora (particularly pollinators and dispersal agents), specific threats to population(s) (e.g. potential for gene flow between CWR and domestication)

8. *Site management policy*: Site objectives, control of human intervention, allowable sustainable harvesting by local people and general genetic resource exploitation, educational use, application of material transfer agreements

9. *Taxon and site population research recommendations*: Taxon and reserve description, autoecology and synecology, genetic diversity analysis, breeding system, pollination, characterization and evaluation

10. *Prescription (management interventions)*: Details (timing, frequency, duration, etc.) of management interventions, population mapping, impact assessment of target taxon prescriptions on other taxa at the site, staffing requirements and budget, project register

11. *Monitoring and feedback (evaluation of interventions)*: Demographic, ecological and genetic monitoring plan (including methodology and schedule), monitoring data analysis and trend recognition, feedback loops resulting from management and monitoring of the site in the context of the site itself as well as the regional, national and international context

Once the taxonomic, demographic and genetic diversity levels for the specific target taxon or taxa within the reserve are described and details included in the management plan, they form the foundation for future monitoring and assessment of change in diversity (see Iriondo *et al.*, Chapter 4, this volume).

However, changes in taxon, demographic and genetic diversity are a natural characteristic of community dynamics. The management plan must allow for natural fluctuations due to stochastic (severe weather, floods, fire and epidemics), as well as cyclical (density-dependent interactions, which may be dramatic but whose effects do not persist) and successional (directional, which may be halted by management intervention) changes. Stochastic and cyclical changes in the short term may be quite dramatic, but will rarely lead to species extinctions (Hellawell, 1991), although they are likely to lead to genetic drift (see Gillman, 1997). The management plan should also attempt to identify the normal limits of natural change in order to determine a threshold value of acceptable change in the target taxon's population size, sometimes referred to as the limit of acceptable change (LAC). However, the identification of the normal range of population levels may in itself require significant research, unless the target taxon has previously been studied at the site. If the latter is not the case, the normal range will only become apparent after several cycles of monitoring; the greater the number of monitoring cycles, the more accurate will be the estimate of the normal limits of natural change.

If monitoring detects that population size has fallen below the established threshold value, management intervention would be triggered. Having emphasized the natural changes seen in plant populations, humans undoubtedly have the most dramatic effect on communities, through incipient urbanization and pollution, or changes in agricultural and forestry practice, for example. Therefore, the management plan must be sufficiently flexible and pragmatic to accommodate certain changes in the anthropogenic impact on the reserve, while identifying and controlling those changes that would be likely to seriously threaten the target population. For example, if a local indigenous group has limited wild-harvesting rights within a particular quadrant of the buffer zone of the reserve, but the harvested species population numbers in that quadrant are dwindling because of lack of time for regeneration, rather than stop all wild harvesting by the indigenous group they could be permitted to continue wild harvesting in another quadrant with a more healthy harvested species population. However, if the harvesting process is destructive and unsustainable, and following consultation no compromise can be reached with the local indigenous group, it may be necessary to halt wild harvesting to save the wild-harvested species.

Given (1994) stresses that preserving communities is not necessarily the same as preserving genes; Maxted *et al.* (1997b) conclude that it is quite possible to preserve a habitat or community and still lose genetic diversity, if not species with that habitat or community. Therefore, it is vital that the reserve be designed and managed in an appropriate manner to maintain the genetic diversity of the target taxon or taxa, and if this objective is threatened, corrective action must be automatically taken.

The management plan is not a document written when the reserve is initially established and set for all time. It will require regular revision to take account of changes in policy, conservation objectives, site biotic and anthropogenic description and research agenda. Most importantly, the management prescription should be seen as dynamic, changing to meet the target taxa conservation goals as they are better understood and policy context as it evolves.

3.6 Minimum Content of a Genetic Reserve Management Plan

The proposed elements of a genetic reserve management plan summarized in Box 3.1 are comprehensive, but it should be realized that there may be many practical or pragmatic reasons why it is not possible to cover all of these issues in an individual reserve management plan. Much of the taxon and reserve descriptive information, such as the breeding system and minimum viable population size, may be unavailable and the genetic structure of the target taxa will often be unknown. As the time and resources required to carry out the research to provide this information may be limited, it may be necessary to balance local community development aspiration with biological expediency, which may therefore compromise pure conservation goals.

Writing a management plan is not purely an academic exercise; it aims to facilitate site management and target taxa conservation. Therefore, the more detailed the management plan, the more useful it will be. However, where it is not practical to complete all elements of the management plan as envisaged in Box 3.1, it would still be necessary to record the accessible information or that which can be collated with the limited resources available as well as outline the management prescription.

3.7 Design and Implementation of a Genetic Reserve Management Plan

There are many texts or aids available to assist in the design and implementation of a management plan; for example, the Eurosite network has developed a Management Planning Toolkit (Idle, 1997, 2000; Idle and Bines, 2004) available at http://www.eurosite-nature.org/article.php3?id_article=77. But this toolkit, which provides a step-by-step guide to the design of a traditional management plan where the focus is primarily habitat conservation, does not consider the genetic component of genetic reserve conservation. Therefore, it is necessary to adapt these texts or aids to meet the specific requirements of genetic reserve conservation.

As such, once the various competing locations for the reserve site have been assessed and the final reserve location selected, the structure of the reserve can be designed to meet the broad conservation goals for the designated target taxa (Fig. 3.1), the assumption implied here by the use of the plural taxa being that establishing a reserve for a single CWR will not be cost-effective and there will most commonly be a need to describe the multiple CWR taxa present at a site. The next stage in establishing the genetic reserve is the writing of the formal management plan for the genetic reserve (Maxted *et al.*, 1997c). This will involve clarifying the conservation context, collecting available information and possibly researching the abiotic, biotic and anthropogenic characteristics of the site, describing the various target taxa and their populations present in the reserve site and, on the basis of all this information, generating the management prescription which details the required management interventions. Due to the cost of establishing a genetic reserve *de novo*, a second assumption is that most commonly pre-existing protected areas will serve as genetic reserves given the necessary minimum change to their management plans. This will require compatibility of the CWR management goals with all other conservation objectives of the protected areas (i.e. genetic conservation of bamboo CWR along with protection of pandas in the same reserve in China). As such,

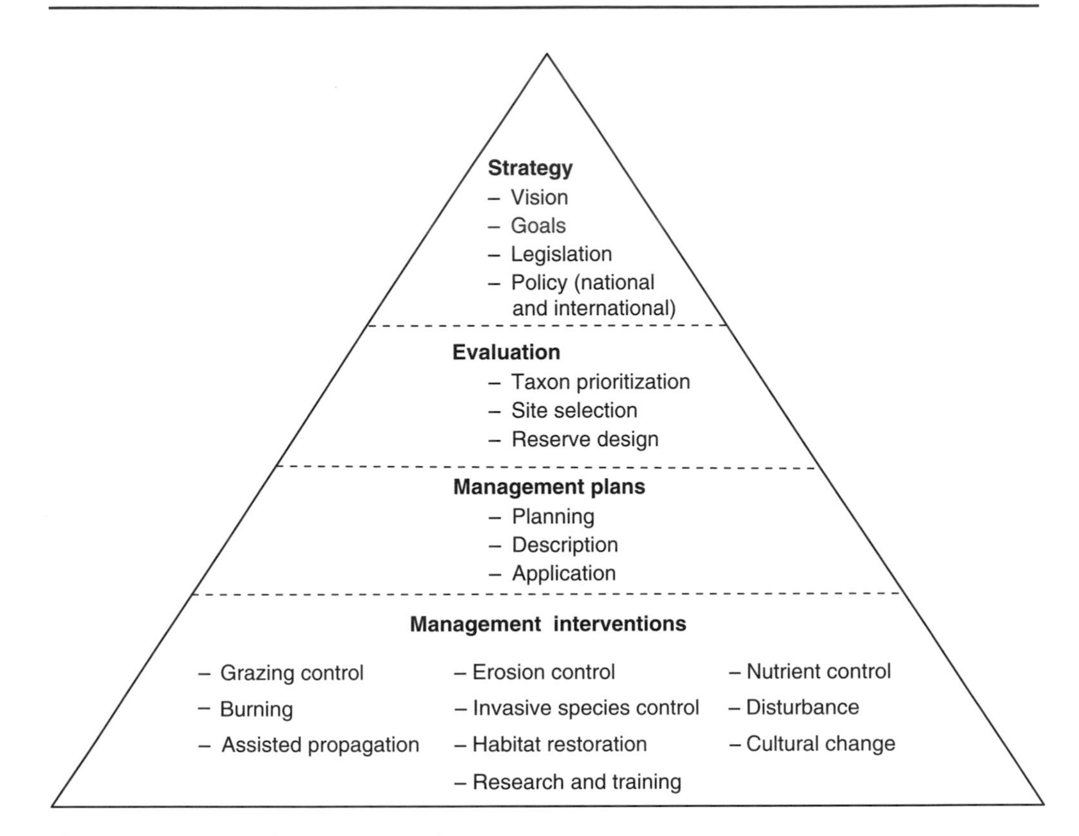

Fig. 3.1. Overview of management planning.

the management plan itself will not be written *de novo* but the establishment of the genetic reserve will mean either the revision of an existing management plan or the writing of a genetic reserve management plan where much of the detail can be taken from the existing protected area management plan.

Although the reserve site will have been selected because it contains abundant and hopefully genetically diverse and/or unique populations of the target taxa, there will be a need to observe and describe the anthropogenic, biotic and abiotic qualities and dynamics of the site to maintain this diversity. Once the ecological dynamics of the reserve are known and understood, a management plan and intervention regime that promotes these elements, as they relate to the target taxon, can be proposed.

The general long-term goal of the genetic reserve is to maintain target taxa diversity and dynamics, and this can only be achieved by having a regime of minimum, effective management interventions, which are detailed in the management prescription. Therefore, the first step in formulating the prescription will be to observe the various dynamics of the site. It should be surveyed so that the species present in the ecosystem are known, the ecological interactions within the reserve understood, a clear conservation goal decided and a means of implementation agreed.

The proccss of writing a genetic reserve management plan is summarized in Fig. 3.2. It can be seen that the process is divided into three phases: planning, description (reserve and target taxon) and management application. Once

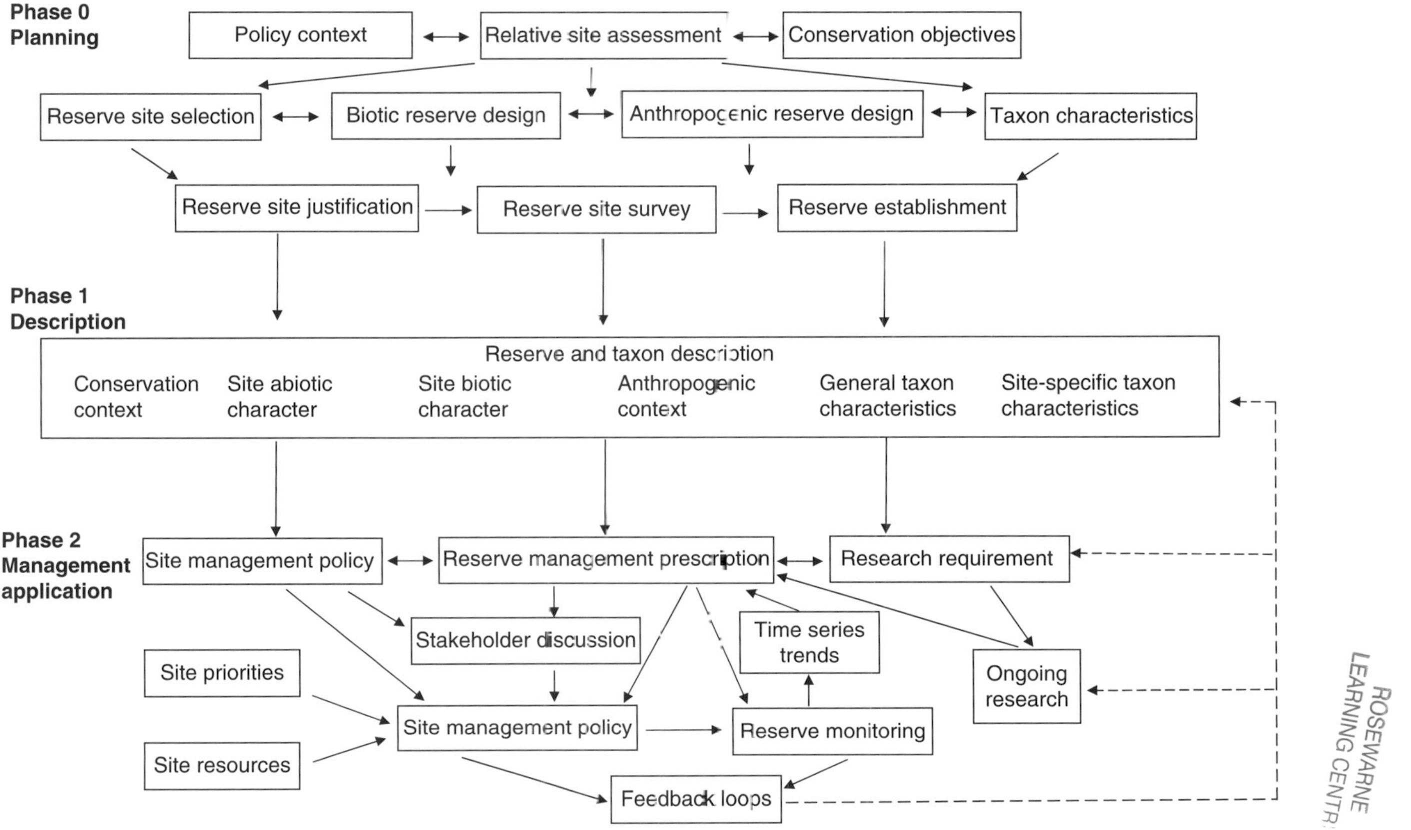

Fig. 3.2. Process of compiling a genetic reserve management plan.

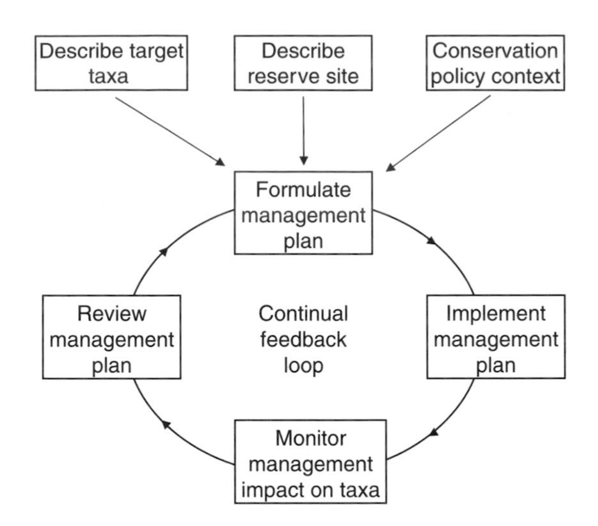

Fig. 3.3. Management plan reviewing.

established, the reserve management plan, and therefore the site management, is a cycle of description and management application being regularly modified by a series of feedback loops (see Fig. 3.3). As such, both the genetic reserve management plan and the management of the site itself can be sensitive to the local biotic, abiotic and anthropogenic changes and are dynamic, adjusting the site management to meet changing management goals.

Wherever CWR genetic reserves are established they will never be established in an anthropogenic vacuum. There are likely to be local farmers, landowners and other members of the local population who utilize the proposed reserve site and who are likely to remain neighbouring communities once the reserve is established. Their traditional use of the site may be disrupted by the establishment of the reserve, but it is also unlikely that any genetic reserve could succeed in the absence of local support and it will definitely fail if the local population opposes the establishment of the reserve. As such, it is important to involve the local community in reserve establishment and the drafting of the management plan. Time should be taken to explain the reasons for the establishment of the reserve to the local community, seek their approval and when designing the reserve take into consideration their needs and aspiration. The local communities' requirements need not conflict with scientific conservation. Traditional utilization (harvesting, hunting and even grazing) is often intrinsically sustainable and even if restricted in the core area could be encouraged in the buffer or transition zones though regulation may be required. There may be a need to compromise between traditional utilization and conservation objectives to ensure success of the reserve (Dudley *et al.*, 2005).

The involvement of local people should not be considered a distraction or disadvantage, since they may be able to assist through volunteer schemes in the routine management and monitoring of the reserve. The personal experience of the authors has indicated that local people are very proud to find that their environs contain 'important' plant species and are very willing to assist in their conservation. This applies to rural communities both in developing and developed countries. It facilitates the development of goodwill between the professional conservationist and local communities neighbouring the reserve. However, in some cases there may be a need to provide other incentives to engender goodwill.

Finally, as pointed out by Morris (1991), 'a plan is no substitute for the management it prescribes'. In other words, there is no point having a beautifully constructed management plan if that plan is not implemented. Associated with this is the failure to recognize the ultimate goal; the plan and management interventions are not a goal in their own right, but merely a means of attaining the conservation goal, a healthy, genetically diverse CWR population. Therefore, a management plan should be a succinct document that identifies the key features or values of the reserve, elucidates the management objectives to be met and indicates the management interventions to be implemented. It also needs to be sufficiently flexible to cater for unforeseen events which might arise during the currency of the plan (Thomas and Middleton, 2003).

Some experiences of writing and implementing management plans in the West Asia centre of plant diversity are discussed in Box 3.2, the text being adapted from Al-Atawneh *et al.* (2007).

Box 3.2. Experiences of writing and implementing management plans in West Asia. (Adapted from Al-Atawneh *et al.*, 2007.)

The West Asia region contains one of the three major mega-centres of crop diversity of global significance, notably the gene pools of wheat, barley, lentil, chickpea, almond, pistachio and pear, and many forage species have their centre of diversity in the region. Due to the sheer number of CWR species present and the lack of resources for extensive *ex situ* collection and conservation, *in situ* conservation is seen as a regional priority. As such, the Dryland Agrobiodiversity Project (focusing on CWR conservation in Jordan, Lebanon, Syria and the Palestinian Territories) targeted the establishment of CWR genetic reserves in two sites within Jordan, Lebanon, Palestine and Syria.

Target taxa were selected depending on the basis of their importance for food and agriculture in the region as potential gene donors; the following gene pools were included: *Medicago* L. spp., *Vicia* L. spp., *Trifolium* L. spp., *Lathyrus* L. spp. and *Lens* Mill. spp.; *Triticum* L. spp., *Avena* L. spp., *Hordeum* L. spp. and *Aegilops* L. spp.; *Amygdalus* L. spp.; *Prunus* L. spp.; *Pyrus* L. spp.; *Pistacia* L. spp. and *Olea* L. spp.; and *Allium* L. spp. The two sites in each country were chosen on the basis of abundance of the target species, representation of diverse ecosystems and absence of major threats to genetic diversity. The areas selected were: Mowaqer and Ajloon in Jordan, Arsal and Balbak in Lebanon, Hefa and Sewaida in Syria, and Hebron and Jenin in the Palestinian Territories. The distribution, frequency and density of the target species were surveyed in these areas annually between 2000 and 2004 and the details of the current site management noted.

The management and monitoring of sites became the responsibility of the Ministries of Agriculture but site ownership was largely private, so it was essential that the management plan for the reserve was developed and implemented in partnership with landowners. This necessitated a public awareness programme and the development of incentives for farmers to maintain CWR diversity and avoid overgrazing, species replacement and changes in land management. For example, in the Wadi Sweid Reserve in Lebanon, which contains many fruit tree CWR species, the cultivated lands were given increased protection from grazing by the appointment of local guards (Natours) who regulated grazing at the community level. While in the Wadi Sair reserve in Palestine, large CWR populations are traditionally protected as they grow on the margins of fruit tree orchards that are traditionally protected from animal grazing by local custom, even though elsewhere populations are significantly smaller due to severe overgrazing.

One major complication encountered by the project was the complexity over the mosaic of land tenure in the West Asia region, as many of the 'best' sites are privately owned with a

Continued

Box 3.2. *Continued*

long tradition of open grazing, medicinal plants collection, herbage or tree cutting and even land reclamation to introduce new crop species. Experience showed that the smaller the number of owners, the easier it was to reach an agreement. Local institutions and local farmers' bodies were involved in discussing the management plans, so that municipality as a whole was consulted. Thus, the community is a real partner in the conservation planning and action. Broader national NGOs were also consulted to facilitate broader national support. Thus, the introduction of a management plan to control these activities involved extensive work with, and education of, local communities to ensure that the implementation of the management plan went smoothly. The management options were selected and discussed with local communities and the owners of the areas to be protected in order to combine conservation aspects with sustainable use and ensure the benefits to the local users. The benefits included technical packages (e.g. road improvement, specialist school teaching), institutional arrangements (e.g. establishment of cooperatives to aid product sales), value added (e.g. selling jam rather than fruits orgrafting stocks) and alternative sources of income that can enhance the livelihood of local communities and so preserve local agrobiodiversity and all the ecosystem functions.

Although all management plans are unique, even if they use as a basis the structure outlined in this chapter, due to target taxon, area, resources and staff availability, etc., there are differences. Particular differences associated with those developed in West Asia are, for example, the necessity for reseeding and replanting with native species because of the general level of overgrazing in the region, the application of water harvesting to promote seeding, the introduction of alternative feed resources for farmers at key times of the year but most importantly the control of livestock grazing. Within the plans among the options for improving incomes for local communities are improved dairy and apicultural production, assisted cultivation of medicinal plants and ecotourism, each implemented in collaboration with local NGOs and lead farmers. Also for the most economically important and highly threatened species, such as *Triticum* L. and *Lens* Mill., the establishment of new protected areas to conserve the remaining populations was proposed.

At each site a monitoring schedule has been established and the Ministry of Agriculture, following discussion with the Economic Development, Education, Environment, Finance, Forestry, Rangeland, Rural Development, Tourism, Trade, Transport and Water Ministries, is responsible for collating and analysing the time series data. The staff are using stratified random monitoring: three transects of 200 m for each site. Five quadrats (plots) of 1 × 1 m are taken along each transect with at least 25 m between each quadrat. Thus, 15 plots were recorded for the herbaceous species inventory at each site. For each quadrat the following features were recorded: species name (scientific and Arabic), family, cover, frequency, abundance, phenology and vitality. For the fruit trees in each of the project sites two transects were placed along the slope gradient and five quadrats of 20 × 20 m were placed on each transect. For each quadrat the following features were surveyed in relation to species information: species name (scientific and Arabic), family, cover, height, perimeter (circumference of the tree trunk) at 50 cm, age, phenology and vitality. The management and even the monitoring of the site will be modified according to the results of monitoring.

The proposed management plans differed from the ideal; as might be expected there is variation between the theory and the practice. Specifically this variation was primarily associated with the need to integrate CWR with landraces conservation and industrial agriculture, and the challenges associated with locating reserves on privately owned rather than governmental land. Compromise and flexibility were key terms in developing the plan. Certainly the establishment of the reserves would have been more straightforward if the reserve had been sited on governmental lands. However, it could be argued that locating genetic reserve primarily on private-owned land did mean that the project was forced to pay more than just lip service to working with the local communities and local authorities, as well as ensuring that the local communities really understood the importance to them of *in situ* agrobiodiversity conservation.

3.8 Management Interventions

Close observation of the target taxa diversity and dynamics along with the observation of the species with which the target taxon interacts will highlight a series of thresholds for management interventions at the site. For example, for many sites where there are 'healthy' CWR populations, the site is likely to have had a certain intensity and timing of grazing and under this specific regime the target taxa have historically thrived. Once the site is designated as a genetic reserve, if the target taxa populations are to remain healthy, the desirable level of grazing needs to be maintained. The management intervention at the site, at least as regards grazing, would be restricted to monitoring the intensity and timing of grazing and ensuring that the regime was not deleteriously changed. If the monitoring showed that the regime was changing and that this change was having a deleterious impact on the target taxon populations, more active intervention would be required to restore the beneficial grazing regime (Plate 6). Some common forms of management interventions are described as follows.

Nutrient control: Many natural ecosystems are nutrient-poor, certainly by comparison to cultivated lands. Therefore, changing the natural nutrient composition of a site will have a dramatic effect on the natural flora, just as eutrophication has a dramatic effect on waterways. Within the reserve areas, artificial nutrient enrichment may often be spotted by the presence of alien species. The intervention required is identification of the nutrient source and prevention of further pollution. However, this process may be complex because of remote pollutant emissions and long-distance dispersal via wind patterns.

Erosion control: In many areas of the world, water and wind erosion of the site can seriously impact on the target species and their populations, altering the physical, topographical and edaphic conditions of the habitat, possibly making the site less favourable for the CWR species. In order to stop or minimize the destructive effects of erosion on the habitat and vegetation, simple and cheap structures such as small dykes, small dams, stone or brush bands, pits, water spreaders, and windbreaks may be established in the reserve and its surrounding areas. A dense and vigourously growing vegetation is the best and the cheapest way of controlling erosion. The soil with vegetation can best be protected against erosion and maximum amount of rainfall may be saved in place by a dense plant cover. However, care must be taken when selecting species for this purpose as several non-native species used for erosion control have proved to be invasive and so lead to another set of problems.

Burning: Fire is a natural part of the ecosystem cycle in Mediterranean environments. Although plants are burnt and killed, the reduced vegetation biomass means that there is less competition for light, nutrients and water for any surviving individuals, new seedlings or opportunist species and so ultimately the longer-term effect of fire may be beneficial to the ecosystem. Most of Australian biota are adapted to periodic fires (Cropper, 1993) and the production of ethylene during the fire stimulates flowering in several monocotyledonous species (e.g. *Xanthorrhoea* spp.). Increasingly, fire is managed to avoid the associated damage to human activities, but managing fire is difficult: too often, and the vegetation may not have time to recover from the previous burning; too rare, and the brushwood build-up may lead to an overly intense fire sterilizing the site. The appropriate fire regime can only be known from careful synecological studies where different blocks of vegetation are subject to different fire regimes, though autecological studies of fire on the local keystone species can provide a rough estimate of the fire regime. If an area is close to human habitation, fire may not be an option, thus ensuring a firebreak around urban areas is necessary and vegetation may need to be controlled by regular slashing and removal of the cut vegetation.

Invasive species control: Any plant or animal species found outside of its natural distribution are referred to as alien species, but not all these species are invasive and directly or indirectly have a deleterious effect on native species. Invasive species are those found outside of their natural distributional range as a result of human intervention and which have become established in natural or semi-natural habitats; they act as agents of habitat change and threaten native biodiversity. These plant or animal species can replace diverse native ecosystems with single, exotic species stands, compete with native species for resources and alter the fire regimes, soil chemistry, geomorphological processes and hydrology of the host habitat, or simply predate native species, all of which can be to the detriment of CWR taxa. The precise method of control will depend on the invasive species, but measures can be broadly grouped into physical, chemical and biological. However, when applying any control measure it is important that the control is less harmful to the native biodiversity than the invasive species itself.

- Physical control of invasive species – This includes a range of methods including hand-pulling of herbaceous weeds or cutting and slashing of shrubs and trees, and is often associated with voluntary conservation groups. This work is strenuous, time-consuming and often impossibly expensive for professional organizations to undertake. However, if undertaken, care must be taken to remove all plant parts allowing the species no chance of regrowth. Burning may be an option in certain habitats, where the burning itself will not permanently harm the native biodiversity and will control the alien. The culling of invasive animal species is often necessary but can be controversial, especially the culling of large mammals, but these are a degree of magnitude easier to control than say rats or bruchid species. The conservationist should be aware, however, that overhasty eradication of alien species can leave empty niches that are potentially open to even worse invasive species, so post-clearance of the site must be monitored and managed to ensure native species' re-establishment. Physical eradication of alien species alone will rarely control their spread; the examples that work well are usually associated with control of invasive species on relatively small islands, such as the eradication of rats from Round Island and Ile aux Aigrettes, off Mauritius in the Indian Ocean (Maxted *et al.*, 1997c), or the eradication of *Carpobrotus* in Minorca (Fraga *et al.*, 2006). The conservationist just has to miss a few individuals for the alien species to survive and reinvade.
- Chemical control of invasive species – The application of chemicals to halt invasive species is not as straightforward as it might appear at first. There may be a need for expensive repeated chemical application to ensure complete eradication and halt reinvasion. Many herbicides or pesticides are non-specific, and so will impact equally on the native species and alien species. Chemicals

may also have residual, long-term effects on the environment and may be just as deleterious as the alien species themselves.

- Biological control of invasive species – The deliberate introduction of an alien species' natural enemies from within its native range can halt their spread. This control method can be inexpensive and efficient, but care has to be taken with the introduction of another alien species to the habitat to ensure that the biological control species is sufficiently focused on the target alien species so that it does not become an invasive species in its own right. This requires an extensive experiment phase prior to widespread introduction of the biological control species. There are many examples where biological control has succeeded but also cases of failure. Following the success in Australia of the introduction of the moth *Cactoblastis* to control cacti spread, it was introduced to Florida, USA, but here it also attacked the native *Opuntia* species. However, biological control can be alien-species specific, as is the case for many invertebrate biological control species; it is not labour- or resource-intensive and has no residual ecosystem toxicity effect.

In reality the effective conservation manager has to be pragmatic, balancing the threat of invasive species against the cost of control and their effect on the CWR species. Often a combination of physical, chemical and biological control measures is required, but most effective of all is prevention of introduction of the invasive species through public awareness of the problem and sound overall reserve management.

Disturbance: Protected areas, particularly smaller reserves, are by definition artificial in a biological sense as loss of diversity is not always balanced by immigration. Therefore, the management regime may necessarily include habitat disturbance, which results in the desired patchwork of diverse habitat types that favours the target CWR species. Natural causes of disturbance include fires, storm damage, pest and disease epidemics, herbivory, floods and droughts. All of these factors are non-uniform, in terms of coverage, and create habitat patches of an earlier successional stage, which will, in turn, promote species and genetic diversity. Within a larger reserve there will be a natural, dynamic patchwork of differing successional stages and the species associated with each stage will migrate between patches. Pickett and Thompson (1978) refer to minimum dynamic area, which is the smallest area with a complete, natural disturbance regime. This area would maintain internal recolonization to balance natural extinctions. As the majority of species are not exclusive to one habitat, the maintenance of reserve heterogeneity will promote the health (genetic diversity) of the full gene pool as represented in multiple populations or metapopulations.

Grazing control: The intensity and timing of grazing is one of the most important management interventions for CWR species, as so many are found in pre-climax communities. While excessive, heavy grazing may destroy a population completely by unsustainable incidental take and disrupting the regeneration activities, it often also results in soil compaction through trampling and the dominance of unpalatable species. Therefore, implementation of sound grazing management is vital for genetic reserve conservation of CWR.

If when the genetic reserve is initially established the target taxa populations are suboptimal, there may be an initial need for complete or partial destocking to facilitate natural regeneration. However, care must be taken regarding when to reintroduce stock. If left too long, vegetation succession may have progressed and the target taxa may be outcompeted for resources by more vigorous shrubs or excessive litter accumulation preventing natural target taxa regeneration (Bakir, 1998), while too early reintroduction of stock may not give sufficient time for natural processes of regeneration.

An essential part of grazing management will be to identify the livestock carrying capacity, periodicity of grazing and kind of stock required at the site. This may not be easy data to obtain with natural pastures in so many regions severely overgrazed and where these questions are not addressed. The livestock carrying capacity of the site is the maximum number and kind of animals which can safely be grazed in a certain range area without detriment to the pasture composition. As a rule of thumb, this is the number of grazing animals that will remove about 40% of the current year's forage production. More than 60% removal of the total forage production is considered heavy or overgrazing and over successive years the target species will not be sustained (Bakir, 1998).

For CWR taxa it is likely that grazing should be light during flowering and then higher after seed ripening of the target taxa. This will encourage seed set; the grazing animals will assist seed shattering and promote trampling of seed into the soil. If possible, rotation of grazing over a site will facilitate recovery of target taxa individuals. If the main target taxon which predominates at a site is highly palatable, grazing will tend to increase overall diversity as the stock will tend to focus grazing on the most palatable species.

Habitat restoration: In terms of CWR *in situ* conservation, habitat restoration must present an extreme option, because if a habitat were of such poor quality, it would not normally have been selected for a CWR genetic reserve. Habitat restoration is a very resource-intensive activity and would only be justified for CWR were no alternative sites for a genetic reserve possible, in which case habitat restoration for CWR species will be restricted to rare endemic CWR. These techniques have been most widely used for re-establishing wetlands, which are largely made up of monocultures of fast-growing species. Habitat restoration is very habitat-specific, and therefore where restoration is necessary, specialist texts should be consulted (see Kell *et al.*, Chapter 5, this volume).

Assisted propagation: It may not always be possible to establish a genetic reserve with the ideal population size, and where it is suboptimal below the minimum viable population, assisted propagation techniques may be used to increase population numbers. The techniques employed include: assisted pollination; collection of seed, germination and planting of seedlings (Plate 12); *in vitro* generation of additional plants and plant translocation (see Cropper, 1993 and Kell *et al.*, Chapter 5, this volume). Artificial reseeding of target species is quite possible but it is very difficult to restore the native vegetation, so even following reseeding the project may fail because of the lack of biotic interactions with associated species. However, artificial reseeding can be successful where the seed to be multiplied is collected and bulked locally, possibly in the reserve buffer or transition zones. There may be a reason why the target taxon is suboptimal at the site; this should be investigated and understood before resources are expended on assisted propagation. If, for example, a species is insect-pollinated and the insect species is in decline, assisted propagation of the target taxon alone will be ineffective.

Cultural change: As emphasized earlier in the chapter, a reserve is likely to fail to meet its conservation objective if it does not have the support of the local community; therefore, it may be necessary to form a compromise between purely scientific objectives and local people's aspirations. The compromise should not only be on the conservation side; local people may also be persuaded to change cultural practices to sit better alongside conservation goals. Hunter-gatherers may be persuaded to avoid collecting rarer target taxa, pastoralists may be persuaded to graze certain areas of the range in a specific order to facilitate flowering and fruiting, or foresters may be persuaded to plant native rather than exotic species. It should be emphasized that when attempting to persuade local people to change their cultural practices careful explanation and sensitivity is critical.

Research and training: Although these are not strictly speaking management interventions, all of the interventions described above can have negative as well as positive outcomes on the genetic reserve if not implemented in the most appropriate manner for the site or CWR taxon. Therefore, research will be needed to better understand the reserve's management requirements. In this sense, every genetic reserve can be considered an experimental area open to routine research related to *in situ* conservation of plants. The success of *in situ* conservation will largely be dependent on the research ability and training of the reserve personnel. In many reserves, particularly in biodiversity hot spots or the centres of crop diversity, there often remains a lack of well-trained staff which is inhibiting successful management of the reserve. There is a requirement for staff to attend entry-level short-term and longer-term international courses to gain the appropriate skills. Likewise, international workshops and symposiums may be organized for professionals to discuss and establish the fundamentals and general principles of *in situ* conservation (e.g. Horovitz and Feldman, 1991; Valdés *et al*., 1997; Zencirci *et al*., 1998; Maxted *et al*., 2007).

Lastly, when considering management interventions, the views of the local community must be considered prior to implementation. For example, in British woodland a common invasive species is *Rhododendron ponticum*, which often forms dense monospecific stands and one method of eradication is to burn the site and control regeneration to favour native species. However, the general public like the showy *Rhododendron* flowers in spring and may be very resistant to eradication. An even more prescient example would be a reserve where fire is a natural causal agent of habitat disturbance and heterogeneity, the burning promoting the target taxon population's health. Reserve management would have to permit continued use of fire, under the instigation of the reserve manager, but if the reserve is adjacent to human habitation, regular fires may be undesirable or even dangerous. As such, novel management interventions should be discussed and agreed with the local community, although the scientifically formulated management intervention may be obvious to the conservation to practically resolve a specific management problem, but again this is another occasion where the views of local people must be considered and possibly compromise interventions agreed.

3.9 Genetic Management Outside of Protected Areas

Many CWR are commonly found in disturbed, pre-climax plant communities (Maxted *et al*., 1997a) and as such many may be excluded from, or marginalized in, established protected areas, which more often aim to conserve pristine habitats, ecosystems or landscapes, or animal species that are now restricted to these environments. Therefore, the genetic conservation of CWR outside of areas conventionally considered protected and designated for conservation must be addressed.

These areas outside of conventional protected area networks where CWR thrive may include roadsides, field margins, orchards and even fields managed using traditional agrosilvicultural practices (Plate 5). In each case, these sites are not managed for biodiversity conservation and the occurrence of CWR populations is purely incidental. However, they often contain large thriving populations of CWR which may be occasionally sampled for *ex situ* conservation but are largely ignored in terms of *in situ* conservation. If these sites are to be considered suitable for sustainable *in situ* conservation, the management they currently receive and that has permitted the existence of a healthy CWR population must be consistent. It might be argued that these sites are more vulnerable to sudden, radical change. A radical change in management would be less likely in protected areas because the *raison d'état* is already conservation, so any management change would more likely be conservative.

Examples of the additional threats faced by non-protected area sites would include: the widening of roads, the scrubbing out of hedgerows or orchards, the introduction of herbicides rather than physical weed control or even the physical control of weeds earlier in the season. Therefore, there is a need to establish some level of protection for these sites, otherwise consistency of management or conservation will be unsustainable. It would be essential to reach a management agreement with the non-conventional protected area site owner and/or manager to ensure that current site management is not radically changed and CWR diversity adversely affected. The management agreement will need to be predicated on an understanding of the conservation context, site characteristics, the target taxon population and the existing management practices that have facilitated a viable population that can be formalized into a site prescription. The prescription will then form the basis of the management agreement between the conservation agency and the landowner. Examples of this form of agreement and prescription are now commonplace in many North American and European countries along rural roadsides, but there are no known agreements yet in place in the centres of CWR where *in situ* conservation of population is a priority. A well-documented example of these kinds of local management agreements are those used in the establishment of micro-reserves in the Valencia region of Spain (see Laguna, 1999; Serra *et al.*, 2004).

Many CWR species are also found growing as weeds in agricultural, horticultural and silvicultural systems, particularly those associated with traditional cultural practices or marginal environments. In many areas of the world this group of weedy CWR species is particularly threatened because of the widespread abandonment of these traditional cultivation systems. Several national governments in developed countries are responding by providing incentives or even financial subsidies to maintain these systems, at least partially to secure continued cultivation and through cultivation to maintain the wild and CWR species that thrive in such habitats. However, the provision of government incentives must be linked to some form of guarantee from the landowner to ensure that wild and CWR species thrive, so again a management agreement including a conservation prescription is required. However, the provision of such grants is unlikely to be a practical option in many developing countries where CWR diversity is largely located but where resources are more limited.

As a specific example of CWR conservation outside of protected areas, the Dryland Agrobiodiversity Project in West Asia found that many intensively cultivated areas contain significant CWR diversity at their margins in field edges, habitat patches or roadsides (Al-Atawneh *et al.*, 2007). Specifically in the base of the Beqaa Valley in Lebanon (Plate 5) which is industrially cultivated there are globally significant populations of rare CWR found along the roadside, while a similar picture was in the Hebron area of Palestine (Plate 2) and in Jabal Al-Druze in Syria where very rare wheat, barley, lentil, pea and bean CWR are common in modern apple orchards. In fact, Al-Atawneh *et al.* (2007) noted that in Palestine, *Pyrus syriaca* Boiss. is only found as scattered trees and never as continuous populations, and so is primarily conserved outside of the existing protected area network. In the latter case, the importance of these isolated trees was drawn to the attention of the local community by use of specific leaflets designed to help raise awareness of this resource, and individual trees were mapped using a GIS so that their long-term presence was easier to monitor.

As by definition the areas outside protected areas are primarily managed for reasons other than conservation, the management interventions at the site are likely to be minimal. The management may in fact be simply maintaining the current management and agreeing not to make radical changes to the site management without discussion with the overseeing conservation officer. The latter will need, however, to routinely monitor the site to ensure that the site management is actually maintaining the target CWR populations.

As such, it is important to emphasize that conservation of CWR is just as feasible outside of conventional reserves as it is within fully designated genetic reserves; a site does not need a fence round it and a sign saying it is a protected area to conserve CWR species to actually conserve CWR species. However, both within and outside of a protected area it is important to have a management plan to ensure that CWR are sustainably conserved. Sustainability is central to CWR conservation and lack of a management plan and management agreement is likely to impede the sustainability of non-protected area conservation. It should also be recognized that there are advantages and disadvantages to CWR conservation outside of specifically nominated protected areas. Due to high levels of resource investment required to establish a more formal genetic reserve, they are in the long term likely to be more sustainable, because to abandon them would be to have wasted the resources already committed to the site. With the lesser commitment of resources, conservation outside of protected areas is more likely to suffer from changes in national or conservation agency policy or changes of ownership of the area where CWR thrive. However, protected roadside or traditional farms act as corridors for CWR gene flow and dispersal and reservoirs to bolster genetic reserve populations. Therefore, to conclude, the effective CWR conservation strategy should include a mixture of conservation within and outside of protected areas to ensure comprehensive and complementary CWR conservation.

References

Al-Atawneh, N., Amri, A., Assi, R. and Maxted, N. (2007) Management plans for promoting *in situ* conservation of local agrobiodiversity in the West Asia centre of plant diversity. In: Maxted, N., Ford-Lloyd, B.V., Kell, S.P., Iriondo, J., Dulloo, E. and Turok, J. (eds) *Crop Wild Relative Conservation and Use*. CAB International, Wallingford, UK, pp. 338–361.

Bakir, Ö. (1998) Management systems for *in situ* conservation of plants. In: Zencirci, N., Kaya, Z., Anikster, Y. and Adams, W.T. (eds) *The Proceedings of International Symposium on* In Situ *Conservation of Plant Diversity*. Central Research Institute for Field Crops, Ankara, Turkey, pp. 67–72.

Cropper, S.C. (1993) *Management of Endangered Plants*. CSIRO, Melbourne, Australia.

De Hond, L. and Iriondo, J.M. (2004) *Population Management Methodologies: Report of Workshop 4*. European Crop Wild Relative Diversity Assessment and Conservation Forum. Available at: www.pgrforum.org.uk.

Dudley, N., Mulongoy, K.J., Cohen, S., Stolten, S., Barber, C.V. and Gidda, S.B. (2005) *Towards Effective Protected Area Systems. An Action Guide to Implement the Convention on Biological Diversity Programme of Work on Protected Areas*. CBD Technical Series no. 18, 108 pages. Secretariat of the Convention on Biological Diversity, Montreal, Canada.

Fraga, P., Estaún, I., Olives, J., Da Cunha, G., Alarcón, A., Cots, R., Juaneda, J. and Riudavets, X. (2006) *Eradication of Carpobrotus (L.) N. E. Br. in Minorca*, IUCN Mediterranean Office. Available at: http://www.iucn.org/places/medoffice/invasive_species/case_studies/eradication_carpobrotus_minorca.pdf.

Gillman, M. (1997) Plant population ecology. In: Maxted, N., Ford-Lloyd, B.V. and Hawkes, J.G. (eds) *Plant Genetic Conservation: The* In Situ *Approach*. Chapman & Hall, London, pp. 114–131.

Given, D.R. (1994) *Principles and Practice of Plant Conservation*. Chapman & Hall, London.

Hellawell, J.M. (1991) Development of a rationale for monitoring. In: Goldsmith, F.B. (ed.) *Monitoring for Conservation and Ecology*. Chapman & Hall, London, pp. 1–14.

Horovitz, A. and Feldman, M. (eds) (1991) International workshop on the dynamic *in situ* conservation of wild relatives of major cultivated plants. *Israel Journal of Botany* 40, 509–519.

Idle, E.T. (1997) *Management Planning Toolkit*. Available at: http://www.eurosite-nature.org/.

Idle, E.T. (2000) *New Guidance for Site Managers: Management Planning Guidance*. Available at: http://www.eurosite-nature.org/.

Idle, E.T. and Bines, T.J.H. (2004) *Complementary Guidance: A Handbook for Practitioners*. Available at: http://www.eurosite-nature.org/.

IUCN (1993) *The Convention on Biological Diversity: An Explanatory Guide*. Prepared by the IUCN Environmental Law Centre, Draft Text, Bonn, Germany.

Karagöz, A. (1998) *In situ* conservation of plant genetic resources in the Ceyanpinar State Farm. In: Zencirci, N., Kaya, Z., Anikster, Y. and Adams, W.T. (eds) *The Proceedings of International Symposium on* In Situ *Conservation of Plant Diversity*. Central Research Institute for Field Crops, Ankara, Turkey, pp. 87–91.

Krebs, C.J. (2001) *Ecology: The Experimental Analysis of Distribution and Abundance,* 5th edn. Benjamin Cummings, San Francisco, California.

Laguna, E. (1999) The plant micro-reserves programme in the region of Valencia, Spain. In: Synge, H. and Ackroyd, J. (eds) *Second European Conference on the Conservation of Wild Plants. Proceedings Planta Europa 1998*. The Swedish Threatened Species Unit and Plantlife, Uppsala, Sweden /London. pp. 181–185.

Lefèvre, F., Barsoum, N., Heinze, B., Kajba, D., Rotach, P., de Vries, S.M.G. and Turok, J. (2001) *Technical Bulletin:* In Situ *Conservation of* Populus Nigra. International

Plant Genetic Resources Institute, Rome, Italy.

Maxted, N., van Slageren, M.W. and Rihan, J. (1995) Ecogeographic surveys. In: Guarino, L., Ramanatha Rao, V. and Reid, R. (eds) *Collecting Plant Genetic Diversity: Technical Guidelines*. CAB International, Wallingford, UK, pp. 255–286.

Maxted, N., Ford-Lloyd, B.V. and Hawkes, J.G. (1997a) Complementary conservation strategies. In: Maxted, N., Ford-Lloyd, B.V. and Hawkes, J.G. (eds) *Plant Genetic Conservation: The* In Situ *Approach*. Chapman & Hall, London, pp. 20–55.

Maxted, N., Hawkes, J.G., Ford-Lloyd, B.V. and Williams, J.T. (1997b) A practical model for *in situ* genetic conservation. In: Maxted, N., Ford-Lloyd, B.V. and Hawkes, J.G. (eds) *Plant Genetic Conservation: The* In Situ *Approach*. Chapman & Hall, London, pp. 545–592.

Maxted, N., Guarino, L. and Dulloo, M.E. (1997c) Management and monitoring. In: Maxted, N., Ford-Lloyd, B.V. and Hawkes, J.G. (eds) *Plant Genetic Conservation: The* In Situ *Approach*. Chapman & Hall, London, pp. 231–258.

Maxted, N., Ford-Lloyd, B.V., Kell, S.P., Iriondo, J., Dulloo, E. and Turok, J. (2007) *Crop Wild Relative Conservation and Use*. CAB International, Wallingford, UK, pp. 1–663.

Morris, M.G. (1991) The management of reserves and protected areas. In: Spellerberg, I.F., Goldsmith, F.B. and Morris, M.G. (eds) *The Scientific Management of Temperate Communities for Conservation*. Blackwell, Oxford, pp. 323–347.

Pickett, S.T.A. and Thompson, J.N. (1978) Patch dynamics and nature reserves. *Biological Conservation* 13, 27–37.

Serra, L., Pérez-Rovira, P., Deltoro, V.I., Fabregat, C., Laguna, E. and Pérez-Botella, J. (2004) Distribution, status and conservation of rare relict plant species in the Valencian community. *Bocconea* 16(2), 857–863.

Thomas, L. and Middleton, J. (2003) *Guidelines for Management Planning of Protected Areas*. IUCN Gland, Switzerland/Cambridge.

Valdés, B., Heywood, V.H., Raimondo, F.M. and Zohary, D. (eds) (1997) Proceedings of the workshops on conservation of wild relatives of European cultivated plants. *Bocconea* 7, 1–479.

Zencirci, N., Kaya, Z., Anikster, Y. and Adams, W (eds) (1998) *Proceedings of International Symposium on In Situ Conservation of Plant Genetic Diversity*. Central Research Institute for Field Crops, Ankara, Turkey.

4 Plant Population Monitoring Methodologies for the *In Situ* Genetic Conservation of CWR

J.M. Iriondo,[1] B. Ford-Lloyd,[2] L. De Hond,[1] S.P. Kell,[2] F. Lefèvre,[3] H. Korpelainen[4] and A. Lane[5]

[1]*Area de Biodiversidad y Conservación, ESCET, Universidad Rey Juan Carlos, Madrid, Spain;* [2]*School of Biosciences, University of Birmingham, Edgbaston, Birmingham, UK;* [3]*INRA, URFM, Unité de Recherches Forestières Méditerranéennes (UR629) Domaine Saint Paul, Site Agroparc, Avignon Cedex 9, France;* [4]*Department of Applied Biology, University of Helsinki, Finland;* [5]*Bioversity International, Rome, Italy*

4.1 Introduction

4.1.1 Definition and objectives of monitoring

Plant population monitoring can be defined as the systematic collection of data over time to detect changes in relevant plant population or habitat attributes, to determine the direction of those changes and to measure their magnitude. Monitoring is an essential component of adaptive management because it improves the knowledge base, measures progress towards meeting the objectives and provides support for maintaining or modifying current management practices (Ringold *et al.*, 1996). In other words, monitoring uses sequential data on the system under study to evaluate effectiveness of management strategies and to make informed decisions about future actions.

The essence of any *in situ* plant genetic conservation project is found in its objectives. The management plan is designed to meet these objectives and monitoring must determine if these objectives are met. For *in situ* genetic conservation of crop wild relative (CWR) populations, the objectives will normally relate to the maintenance of the initial levels of genetic diversity in the target populations and the assurance of the viability of the populations from a demographic, genetic and ecological perspective. In some particular cases where the original status of the populations being conserved is not at its optimum, or when the population has experienced a catastrophic event, the objectives may concentrate on achieving specific targets regarding population size, structure and genetic diversity.

The monitoring of CWR populations and their habitat will have its own specific objectives, which will involve detecting change in certain parameters over time. Some examples of these objectives include:

1. To provide data for modelling population trends;
2. To assess trends in population size and structure;
3. To detect a specific change in population size and structure that will indicate a demographically unstable population;
4. To assess trends in population genetic diversity;
5. To determine the effect of altering or eliminating habitat disturbance (i.e. management actions) on CWR populations.

The value of monitoring in a management context is that detected changes in CWR populations or habitats can provide feedback for timely population management or changes in land use. This feedback also allows genetic reserve managers to direct further research to answer specific hypotheses about the possible causes of population or habitat change. Thus, monitoring results should be analysed regularly as new results are produced. In the cyclical pattern of monitoring, data analysis and management actions (Fig. 4.1), monitoring can be considered an 'early warning' system for management (Bonham *et al.*, 2001).

It is important at this point to make a clear distinction between monitoring and research. While monitoring detects changes in parameters, research is normally oriented to determining the cause of some pattern or process observed. Many monitoring programmes intend to determine the response of a plant population to a particular management activity, but in reality few monitoring programmes conclusively identify the cause of the response (Elzinga *et al.*, 1998).

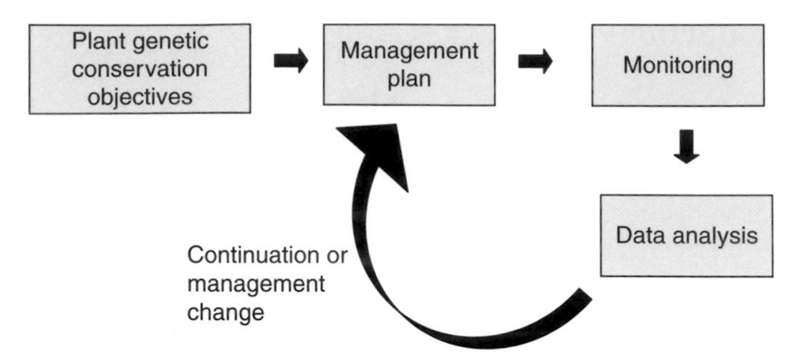

Fig. 4.1. Cyclical pattern of monitoring, data analysis and management actions.

Population monitoring focuses on the target taxon and measures aspects such as population size, structure, frequency, density or cover. Habitat monitoring uses indicators that describe how well the objectives for habitat management are met by the current practices (Elzinga *et al.*, 1998). The latter approach may be of interest when management activities to conserve a population are based on improving habitat conditions for the species. While habitat quality and availability are obviously major factors in the viability of CWR populations, the most direct measure of population trend is provided by demographic data. Habitat monitoring may, however, provide indirect information when demographic data are not readily available (e.g. in *Poa pratensis* or similar grass species detecting changes in population size by making an accurate count of the individuals is very difficult; however, it is quite easy to detect a reduction of herbaceous vegetation in comparison with bushy vegetation in the habitat, which may indicate that the *P. pratensis* population is decreasing). This assumes that the measured change in characteristics of the habitat can be directly related to population change (Bonham *et al.*, 2001).

As well as monitoring the biological status of a target taxon and its habitat, it is equally important to monitor the status of actual or potential threats, which may be biotic, abiotic or social in nature. This perspective is critical to be able to adjust management interventions to minimize or prevent threat impacts.

4.1.2 Initial considerations

Before designing a monitoring programme for a CWR population occurring in a genetic reserve, some basic issues should be considered. A crucial issue in any monitoring programme is the resources available both in the short and the long term. Monitoring resources may depend on international, national and local support, as well as the people and equipment available. If the CWR is a high-priority species, there is likely to be more support and funding for the monitoring programme in the long term. As monitoring involves the systematic collection of data over time, sustainability of the monitoring programme is imperative. Other resources that will influence monitoring design are the personnel available to do the work and their qualifications, access to professionals with specialized skills and available field equipment. When the genetic reserve is located in a protected area, monitoring resources will in theory be easier to attain. Nevertheless, a key point will be the presence of permanent staff working in the protected area. This is essential information, as the types and amounts of resources available will limit the extent and complexity of the monitoring programme.

Another question is the suitability of the taxon- and site-specific information collected during genetic reserve establishment as a baseline or first measurement for a monitoring study. The study carried out for genetic reserve establishment may be inadequate to detect future changes, as the necessary type and intensity of monitoring will often not be known until this initial study is completed. For example, analysis of the initial study may conclude that the design included too many variables or used a sample size that was too small.

The intensity of the monitoring programme will mainly be determined by the priority given to the CWR species in the management plan. Low-intensity qualitative monitoring may be adequate for low-priority or non-threatened species, whereas the rarity of the CWR species, the degree of threats or the political sensitivity of potential decisions may require the use of an intensive demographic approach.

The spatial scale of interest for monitoring should also be identified. Should all populations of the target taxon in the genetic reserve be monitored? Will a single population or part of a single population be enough? It is important to establish the spatial scale in the planning phase, as it will influence later decisions and design. If the scale of interest is a group of genetic reserves established for the conservation of a CWR, it may be necessary to coordinate various reserves to develop a network of monitoring studies.

Before designing the monitoring programme, it is also important to determine if monitoring will be used to detect a percentage change, an absolute change, a target value or a threshold value. Depending on the management objective for the CWR species, it may be necessary to define a desired increase, a critical decrease or a target population size. In any case, the quantity should be measurable (accurately measuring a 5% change in population size may be difficult) and biologically meaningful (a 10% change in recruitment in an annual species is probably not important) (Elzinga *et al.*, 1998).

For large monitoring programmes involving CWR species occurring in more than one genetic reserve or highly controversial species and populations, it may be advisable to assemble a multidisciplinary team to work on the monitoring design.

Efficient monitoring methods for CWR populations or habitats are characterized by stability, power and robustness. Stable methods are those that will not result in false conclusions of change when no change has occurred. Powerful methods are sensitive enough to detect change when it has occurred. Lastly, methods are said to be robust when measurement techniques provide data that are independent of the technique used, e.g. plot size or transect length (Brady *et al.*, 1993).

4.1.3 Monitoring design

Identification and selection of variables that can be monitored

The design of a monitoring programme is based on the specific monitoring objectives previously established. The types of variables to be monitored should be identified and selected in order to answer the questions posed. In CWR conservation, monitoring will require information on demographic, ecological and genetic parameters. Demographic parameters are used to assess the viability of the population. This information can be used to estimate population trends, extinction risks and minimum viable population (MVP) size, as well as to identify the demographic factors that mostly determine population viability. Ecological parameters should be able to identify changes in the physical conditions that operate in the population and characterize the dynamics in the composition of communities associated with target CWR. Finally, genetic parameters may be important for evaluating the genetic diversity contained in the population as well as for understanding the genetic processes that take place in the dynamics of the population. The final goal of CWR conservation is to be able to maintain and provide a source of useful genes that can be applied in crop breeding. Therefore, particular attention must be paid to those processes that lead to genetic erosion and genetic pollution. Thus, changes in breeding patterns, decreases in population size, genetic drift, natural selection and hybridization events could all affect genetic structure and diversity in the population.

Population dynamics, ecological interactions and genetic processes should not be seen in an isolated context since they are very much interrelated. For example, changes in the relative frequency of pollinators often affect breeding patterns. Similarly, an increase in the frequency of inbreeding and loss of heterozygosity may lead to loss of population fitness and inability of the population to respond to environmental change. On the other hand, a decrease in population size, albeit temporary, may activate a process of genetic erosion through genetic drift.

The life history of a plant describes, among other things, the average life expectancy of a plant and how long it takes to reach reproductive stage (Silvertown and Lovett Doust, 1993), whereas the breeding system classifies plants ranging from asexually to sexually reproductive, and in the latter case, from highly outcrossed to moderately or even highly self-fertilizing (Charlesworth and Pannell, 2001). Both the life history and breeding system of the target taxon must always be considered in monitoring design, because they significantly affect demographic and genetic processes, effective population size (see Section 4.3) and distribution of genetic diversity among populations. As a result, the features of a monitoring plan will be different depending on whether the CWR species is annual, biennial or perennial, and a clonal, outcrosser or selfer. Other factors that influence the selection of variables are the morphology of the species and the resources available for monitoring.

Alternative indicators can be used as surrogates for variables that are difficult to measure or monitor (e.g. demographic parameters in very small species, annual plants, dormant cryptophytes or long-lived species). Thus, the breeding system (selfing or outcrossing) may indicate how genetic diversity is partitioned within and among populations. Similarly, current population size or, better, recent change in population size, is an indicator of effective population size, which in turn can serve as an indicator of genetic parameters. Monitoring threat status to indicate habitat and plant population condition can also form an effective basis for management actions.

Census and sampling design

A census of the population counts or measures each individual. The main advantage of this approach is that the measure is an actual count and not an estimate based on sampling (Plate 7). The changes measured from year to year are real, and the only significance of concern is biological.

However, in many situations counting or measuring all individuals of a population is not practical and it is necessary to use sampling methods. The purpose of sampling is to provide information about the total population by gathering data from just part of it, i.e. a subsample. Some key elements of a sampling design include the definition of the sampling unit and the determination of unit size and shape. The sampling unit must be explicitly identified to ensure that the selected units are random and independent. The type and size of sampling unit depend on the variable that is being measured. Thus, individual plants are the sampling units used for attributes such as plant size or number of flowers per plant, whereas plots or quadrats are used for measuring density, frequency or biomass. The most efficient size and shape of the sampling unit depends on the spatial distribution and density of the target population, edge effects, ease in sampling and disturbance effects. A careful selection can reduce the number of sampling units that must be measured, thus reducing the time and resources required for the fieldwork. Krebs (1989) provides equations for determining optimal plot size using Wiegert's or Hendrick's methods based on variation among plots, cost of measuring one plot and cost of locating one additional plot.

The sampling process is usually accompanied by some measure of the quality of the generalization (McCall, 1982). Thus, when monitoring is based on a sampling strategy, monitoring objectives should include specific information such as levels of precision, acceptable levels of power and false-change error rate or minimum detectable change (Elzinga *et al.*, 1998).

When the monitoring objective is to estimate some parameter in a population, the estimate obtained can be compared to a target or threshold value to determine if the management objective is met. In these situations the level of precision should be specified. A confidence interval provides an estimate of precision around a sample mean that specifies the likelihood that the interval includes the true value (Elzinga *et al.*, 1998). For example, the management objective in a genetic reserve of the narrow endemic *Vicia bifoliolata* may be to maintain a population density larger than 0.5 individuals/m^2 over the area of occupation. If a sampling strategy is used in monitoring to estimate population density, the monitoring objective may state that population density is to be estimated with a confidence interval of ± 0.1 individuals/m^2 and a confidence level of 95%.

Consideration of the degree of precision required leads to the issue of how many units should be sampled. The precision of an estimate of a parameter increases with sample size, and the larger the number of sample units, the more precise the estimate will be. However, there is a point above which the estimate becomes only marginally more precise. Information from pilot sampling can be used to determine how many sampling units are required to achieve a particular confidence interval. Specific examples for this calculation can be found in Elzinga *et al.* (2001) and Gibson (2002).

Croy and Dix (1984) calculated the sample sizes necessary to take measurements for some morphological attributes in plants of different life forms. For example, the sample sizes required to bring the 95% confidence interval to within 10% of the mean for plant height were 47, 138 and 135 in *Agropyron smithii* (Poaceae),

Lactuca serriola (Asteraceae) and *Pinus ponderosa* (Pinaceae), respectively. However, the sample sizes required to measure leaf blade length under the same conditions in the same species were 44, 43 and 38. If no preliminary information is available, as a rule of thumb, a minimum of 20 observations are needed to measure a variable with a minimum precision (Dytham, 1999). When dealing simultaneously with various populations, the comparison of variances of traits measured *in situ* among populations should be considered with caution: these values reflect both the environmental and the genetic divergences among populations. Only the most heritable traits could eventually provide some information on genetic diversity. The best alternative in this case is clearly to set common garden experiments.

In other cases, the monitoring objective is to determine whether there has been a change in some population parameter between two or more time periods. When a sampling strategy is followed in this type of monitoring, the acceptable Type I and Type II errors and the desired minimum detectable change must be specified (Elzinga *et al.*, 1998). The Type I error represents the chance of concluding that a change took place when it really did not. It is normally set at $\alpha < 0.05$. The Type II error (power) represents the chance of concluding that a change did not take place when it really did. Calculation of power depends upon sample size, intrinsic variance in the population being sampled and effect size. A description of statistical tests to calculate power can be found in Cohen (1988) and Steidl and Thomas (2001); a review of software suitable for power analysis appears in Thomas and Krebs (1997). The desired minimum detectable change specifies the smallest change that one is hoping to detect with a particular sampling effort. Thus, the monitoring objective of a genetic reserve containing a population of *Narcissus cavanillesii* may be to be able to detect a 20% decrease in density in a 5-year period. The sampling objective may specify that the manager wants to be 90% certain of detecting a 20% decrease in density in a 5-year period, and is willing to accept a 5% chance of making a Type I error. Elzinga *et al.* (2001) provide sample size equations and examples for detecting differences between two time periods.

Why bother specifying precision levels, Type I and Type II error rates and minimum detectable change when designing a monitoring plan? The main advantage is that these specifications are useful for avoiding low-power monitoring studies, which may provide totally misguiding conclusions. It is advisable to consult specialists in statistics or sampling design throughout the design process, so that the monitoring programme provides accurate and meaningful results.

Selection of sampling units

The selection of an appropriate sampling unit and the positioning of the sampling units in the target population are two basic issues that need to be addressed in the monitoring design. The two main ways of sampling vegetation and plant populations are plot methods and transect or intercept methods. Plot sampling involves taking observations within areas of standard size, usually called quadrats. In the line intercept method, a measuring tape is laid out in a random direction at the sampling point and observations are taken on those individuals that intersect the tape. Most monitoring in genetic reserves will probably involve the sampling of temporary or permanent quadrats (Maxted *et al.*, 1997).

Quadrats are usually square or rectangular, although circular quadrats have smaller edge effects. If the quadrat size is small, portable wood or metal frames may be used. Larger quadrats can be demarcated by pacing out or measuring the sides using a tape measure, placing pegs or stakes in each corner, and running string or coloured tape around the perimeter. The required observations are made by systematically going through the quadrat, counting and perhaps measuring and even tagging each individual of the target species encountered (Maxted *et al.*, 1997). As quadrat shape and size can have a major influence on the precision of the estimate, it is necessary to consider the nature of the parameter to be monitored when defining these two features.

Most plant populations are aggregated in their spatial distribution. In this situation, rectangular quadrats yield more precise estimates than square or circular quadrats of the same size. Rectangular quadrats have a greater probability of including some of the clumps of plants inside of them. Consequently, this decreases variation among quadrats and increases the precision of estimates. It is generally best if quadrat length is longer than the mean distance between clumps. Orientation of quadrats can also be important. Rectangular quadrats should be positioned to capture variability within quadrats rather than between them (Elzinga *et al.*, 1998).

Plot size can vary over several orders of magnitude from a few square centimeters to 1 ha depending upon the size and density of the individuals to be studied. A plot size that will result in a mean density below ten is recommended when plant aggregation is high (Hayek and Buzas, 1997). When sampling permanent quadrats it is important to take into account the possible impact of people recording data on the population. Thus, it is best if the quadrats are small enough so that the recorders do not have to stand much within the limits of the quadrat.

Transects are elongated sample areas, either rectangular or lines of zero-width that run through a study site. When estimating cover for clumped populations, the best approach is usually to randomly position transects in the population to be sampled, and, with a random start, to systematically place square or rectangular quadrats of a size that facilitates accurate cover estimation along each transect. The transects, not the quadrats, are treated as sampling units. Line interception can also be used to estimate cover. The theoretical basis of line interception lies on reducing the width of the lines to zero (Lucas and Seber, 1977; DeVries, 1979; Floyd and Anderson, 1987). Because each line is a single sampling unit, the precision of cover estimates will depend on the variation among lines. So, lines should be long enough to cross most of the variability in the vegetation being sampled. When using this method, it is recommended to read along only one edge of the measuring tape and to ensure the tape is not inadvertently moved while sampling. Points can also be used as sampling units when cover is measured with the point intercept method, for example, when monitoring sparse populations of wild fruit trees.

Other types of sampling units that are relevant for monitoring CWR populations in genetic reserves are individual plants (for attributes such as plant height and number of flowers per plant) and plant parts (e.g. fruits if the attribute is the number of seeds per fruit).

Positioning sampling units

When sampling is used to characterize one or more parameters of the CWR target population or its habitat, the question arises of how to position the sampling units in the population. The most important requirement is the use of a random sampling method. The reason for this is that statistical inferences about the whole population cannot be made from a sample unless we apply some type of random selection of sampling units. A second consideration relates to the concept of interspersion. The sampling units must be distributed throughout the whole area of the target population for adequate representation.

Several methods of random sampling are available (Hayek and Buzas, 1997), although the most commonly used are simple random sampling, systematic sampling and stratified random sampling.

SIMPLE RANDOM SAMPLING When this type of sampling is used, each combination of a specified number of sampling units has the same probability of being selected, and the selection of one sampling unit is not influenced by the selection of any other. Only samples generated by a random process have a known pattern that can be used in statistical inference (Hayek and Buzas, 1997). The best way to ensure randomness is to use a table of random numbers or computer-generated random numbers. Random coordinates can then be selected for each of the two axes. The point at which these intersect specifies the location of a sampling unit. Alternatively, the population area can be overlaid with a grid, where cell size is equivalent to the size of each sampling unit. Random coordinates are then chosen for both axes that represent the lower left corner of each quadrat to be sampled. Sampling is done without replacement, i.e. not allowing repetition of a pair of coordinates and rejecting coordinates that fall out of the target population boundaries (Elzinga *et al.*, 1998).

Simple random sampling works well when geographic areas are relatively small and have a homogeneous habitat and the number of sampling units is not very large.

SYSTEMATIC SAMPLING An alternative to simple random sampling, used to avoid practical problems of applying random sampling schemes, is to collect samples at regular intervals, predetermined in either space or time. Systematic sampling is commonly used to position quadrats for frequency sampling and points for cover estimation. This procedure has the advantage of spreading out the samples over the entire target area, thereby providing good interspersion. Systematic sampling is more efficient than simple random sampling, particularly if the area being sampled is large, because of decreased set-up and travel time. This approach can be used in any sampling situation, as long as the first sampling unit is selected randomly and the sampling units are far enough apart to be considered independent. The sampling variance of a systematic sample, i.e. the standard error, depends on the spatial distribution of the individuals in the target population. When spatial distribution is homogeneous or random, the use of systematic sampling is not disadvantageous in terms of variance reduction. However, when the spatial distribution is sparse or patchy, it is not possible to tell whether the estimators for systematic sampling are more, or less, precise than those from random sampling. The use of a replicated systematic sampling approach allows for the estimation of the variance directly from the observed data (Hayek and Buzas, 1997). When the individuals are spread out along a gradient, systematic sampling should take place along that gradient (Hayek and Buzas, 1997). Systematic sampling is undesirable if the pattern of the sampling units intersects some pattern in the environment, e.g. dune ridges and slacks (Goldsmith *et al.*, 1986). If some periodic pattern does exist, the data analysis will not reveal this and estimates, particularly of standard errors, will be wrong (Elzinga *et al.*, 1998).

STRATIFIED RANDOM SAMPLING Stratified random sampling involves dividing the population into two or more groups prior to sampling. Groups are generally defined so that the sampling units within the same group are very similar, while the units between groups are very different. Simple random samples are then taken within each group. These groups should be based on habitat characteristics that are unlikely to change over time, such as soil type, aspect, major vegetation type, and soil moisture (Elzinga *et al.*, 1998). In the context of CWR the grouping can also be based on areas under different management regimes, such as grazed versus ungrazed or cultivated versus uncultivated. Depending on the purpose of the study, sampling units can be distributed evenly among groups, or proportionally to the number of target plants or to the variability in each group.

Stratified random sampling works well when the target population naturally falls into several subdivisions (e.g. when a CWR population extends through several vegetation types). In addition to estimating the overall population value, stratified random sampling provides separate estimates for each group. This feature alone might be reason enough for using this method.

Pilot studies

Once the monitoring study has been designed, a pilot study should be carried out to test the efficiency of the field techniques and assess the experimental design. On evaluating the field data from the pilot study, one may find that the sample size is inadequate for detecting change or that statistical confidence levels are not obtained. At this stage, modifications can be made in the selected methods, saving a lot of time, effort and cost in the long term.

4.1.4 Data recording and monitoring documentation system

A very important part of any monitoring programme is the development of a data recording and documentation system. As some monitoring programmes will undoubtedly include CWR target species occurring in more than one genetic reserve, a consistent and thoroughly described methodology is imperative. The designed monitoring methodology should be clearly documented so that it can be followed by other reserves or by different technical staff within the same reserve.

Data that are not clearly recorded in the field will be of little use when the time comes for data analysis. Data can be collected using field data forms on notebooks, portable computers or personal digital assistants (PDAs). Whichever method is chosen, it is best to predefine codes and to fill in the forms with as much information as possible before going out into the field to avoid repetitive writing and reduce mistakes. Detailed information on the data collection methods defined in the monitoring plan should also be taken along in case doubts arise during collection (Elzinga *et al.*, 1998).

Collected data are normally transferred to a spreadsheet or statistical software for subsequent data analysis. Data gathered as part of a large-scale monitoring network should be stored in a relational database to facilitate data management and data processing (Stafford, 1993). In these situations it is essential that data gathering follows a data structure that is compatible with that of the database it is oriented to. In genetic reserve conservation of CWR it may be very useful to follow the data structure used in Crop Wild Relative Information System (CWRIS) in the framework of the European Crop Wild Relative Diversity Assessment and Conservation Forum (PGR Forum, http://www.pgrforum.org).

4.2 Demographic and Ecological Monitoring Methodologies

4.2.1 Demographic parameters

The main purpose of including demographic parameters in the monitoring of CWR in genetic reserves is to assess the viability of the target population; in other words, to determine if population size is declining or stable and to what extent the population faces extinction. From a management perspective, the information obtained from demographic parameters will be useful for identifying which vital rates most directly influence population viability and for establishing a minimum population size below which specific actions must be taken. Monitoring demographic parameters can also be used as a surrogate for monitoring genetic traits of the population (see Section 4.3).

Population size, density, frequency and cover

Population size is the total number of individuals in a population. For relatively small populations (e.g. below 5000 individuals), population size can be determined through a census, but for larger populations it is more effective to use sampling methods. It is advisable to mark all the individuals that can be found in the population or in the sampling plots with flags or tags and then proceed to count them as the markers are collected. Hand counters can also be very helpful in the counting process. A direct count without any aid will increase the chance of error as some individuals may not be counted at all, whereas others may be counted twice or more.

Density is the number of individuals per unit area. A critical question in measuring both population size and density is the definition of a counting unit which can be consistently recognized by all observers. As opposed to animals, the concept of 'individual' in plants can be confusing, due to their modular architecture and various systems of asexual propagation. From a CWR perspective, we will ideally be interested in counting individuals that correspond to distinct genetic entities. However, in practical terms, it may be necessary to define a counting unit (stem, group of stems) that allows for clear, consistent counting, although this unit does not necessarily correspond to a genet. Both population size and density are most sensitive to changes caused by mortality (death) or recruitment (birth). Therefore, these two parameters may not be adequate measures for long-lived plant species that respond to stress with reduced biomass or cover, rather than mortality, or for CWR populations that fluctuate dramatically in numbers from year to year, such as graminoid annuals.

Two alternative measures of population size are frequency and cover. The main advantage of these measures is that data are quite easy to obtain. Furthermore,

Plate 5

Plate 6

Plate 5 CWR conservation on road side
An alternative approach to establishing a genetic reserve is to manage more informal sites to promote plant genetic diversity conservation. Here conservationists are assessing the management of a roadside and field headland rich in cereal and legume crop wild relatives in the Northern Bekaa valley, Lebanon. (Photo credit: Nigel Maxted)

Plate 6 Giant Aldabra tortoise
Conservation management measures are required to maintain Genetic Reserves in a healthy state, to help control invasive species and to restore the dynamics of the ecosystem. In this illustration, Giant Aldabra tortoises (*Dipsochelys dussumieri* Gray (syn. *Geochelone gigantea* Valvede)) have been introduced to Ile aux Aigrettes Nature Reserve in Mauritius as a surrogate to the extinct land tortoise of the island (*Geochelone inepta* and *G. triserrata*) to maintain a sustainable ecosystem. (Photo credit: Ehsan Dulloo)

Plate 7

Plate 8

Plate 7 Population census
Plant officers of the Valencian plant microreserves programme (Valencian Community, Spain) make a census of the recently discovered *Lupinus mariae-josephi* on a burnt area near Xativa. (Photo credit: Emilio Laguna)

Plate 8 Monitoring
Once a reserve is established, the population dynamics of the targeted species need constant monitoring. This picture shows a monitoring plot for the recently discovered endemic *Lupinus mariae-josephi* in Montserrat (Valencian Community, Spain). The plot corners are marked with biodegradable paint. (Photo credit: Emilio Laguna)

they are especially well suited as monitoring parameters under specific circumstances. We can define frequency as the percentage of plots occupied by the target species within a sampled area (Plate 8). Thus, if a grid of sampling units is laid on the study area, the percentage of those units occupied by the species is the frequency. In this case, the abundance of the species within the plot or sampling unit is irrelevant, as occupation is just a matter of whether the species is present or not. As frequency measures depend on plot size and shape (Hayek and Buzas, 1997), frequency values from studies with different sampling units cannot be compared. This parameter is appropriate for monitoring annual CWR, whose density may vary dramatically from year to year, but whose spatial arrangement of germination remains fairly stable. Rhizomatous species, especially graminoid (*Poaceae*) species growing with similar vegetation, are often measured by frequency because there is no need to define a counting unit as with density (Elzinga *et al.*, 1998). Frequency is also a good measure for monitoring invasive species that may pose a threat to a target CWR population. A clear advantage of this parameter is that it can be measured with minimal training, as the data recorder only needs to determine whether or not the species occurs within the plot.

Cover is the percentage of plot area that falls within the vertical projection of the plants of the target species. The percentage of plot area projected by the whole aerial part of the plants is canopy cover, whereas basal cover is the percentage of plot area covered by the base or trunk of the plants. It is an interesting parameter because it is highly correlated to biomass or annual production and it matches the contribution of species that are very small, but abundant, and species that are very large, but scarce (Elzinga *et al.*, 2001). It is easier to measure the cover of matted plants and shrub species with a well-defined canopy, although cover measurements are applicable for nearly all types of plants. Cover measurements are often used for grasses because of the difficulty in counting tillers. A key benefit of using cover is that it does not require the identification of the individual. Nevertheless, canopy cover can significantly change over the course of a growing season, while both frequency and density are more stable. Furthermore, cover measures are more difficult to make with a high level of precision.

Population structure

For most CWR, particularly those that are long-lived, individuals differ in important ways that affect their current and future contributions to population growth. For example, larger individuals often have a greater chance of surviving and a higher reproductive rate than smaller individuals. As a result, two populations that differ in the proportion of large versus small individuals may have very different viabilities, even if they are the same size and experience the same environmental conditions. Thus, population structure is an important feature to monitor in a population. Populations are usually structured by age, size or stage. One important basis for making this decision is simple practicality. For example, size can often be quickly measured in the field. An ideal structuring variable will be highly correlated with all vital rates of a population, allowing accurate prediction of an individual's reproductive rate, survival and growth. A major drawback of some measures of size is that they do not accurately predict vital rates. For example, the future survival and growth rates of some plants with underground storage organs, such as *Beta vulgaris*, *Brassica napus* and *Daucus carota*, may be only marginally predicted by the current height of the stem, because above-ground height is not well correlated with underground stores of carbohydrate. A second desirable feature in the structuring variable is accuracy of measurement. If repeatability of measurements is low, there will be little accuracy in estimates of growth. Some measures of size, such as total leaf area, may be so difficult to measure accurately, at least with the amount of time needed to measure hundreds of individuals, that they are poor choices as structuring variables. Thus, the best variable is one that achieves a balance between accuracy and practicality (Morris and Doak, 2002).

One of the simplest approaches is to use stage classes such as seedling, non-reproductive and reproductive. Sometimes, in search of ease and maximum efficiency, monitoring can just focus on mature reproductive individuals (e.g. IUCN, 2001). Stage and size criteria are often used simultaneously (Box 4.1). However, age is not normally used as a criterion for population structuring in plants, because it is usually a difficult variable to measure in a non-destructive way and does not generally correlate well to plant vital rates. Whatever the criteria used, population size, density and frequency estimates can be obtained separately for each stage class. Although measuring any of these parameters by stage class increases the amount of time required to evaluate each plot, it also provides clear advantages. For instance, measuring density in stage classes can rectify the insensitivity of density as a measure of some kinds of change, whereas assessing occurrence by stage class can improve our understanding of frequency change.

Box 4.1. Population viability analysis in *Erodium paularense*. (From Albert *et al.*, 2004.)

Erodium paularense Fern. Gonz. & Izco (Geraniaceae) is a woody rosulate chamaephyte, endemic to central Spain. This species has been classified as endangered (EN) (VV.AA., 2000) according to the IUCN categories. In addition to its narrow distribution and the small size of the populations, these plants have very low reproductive success (González-Benito *et al.*, 1995; Albert *et al.*, 2001). Available data also show evidence of seed predation by ants and low seedling recruitment. Furthermore, populations are also subject to human impact, such as cattle herbivory and the effects of recreational activities and plant collection.

Taking into account the plant size and the ability to produce flowers, plants were grouped into four stages, one vegetative and three reproductive. These classes were obtained from field data by cluster classification and comprise the following categories:

Juvenile	<6 cm
Adult I	6–12 cm
Adult II	13–21 cm
Adult III	>21 cm

A PVA was made with data from the smallest population of the Lozoya Valley, with an occupancy area of 443 m². In this study a metapopulation model was implemented with the patch structure of its five subpopulations, four located in rock microhabitat and one in lithosol. The metapopulation model was built using both spatial and demographic information gathered from the two microhabitats since 1993. Different simulations were run to estimate extinction risk and population decline under present and possible future scenarios, and to evaluate the effectiveness of different conservation actions (Table 4.6).

Vital rates

Along with current population size, population viability essentially depends on the vital rates of the population. These are responsible for the births and deaths that determine population growth. Three types of vital rates can be identified:

- Survival rates: The survival rate is the proportion of individuals at one census that are still alive at the next census. Survival rates are usually estimated for each single class in structured populations.
- Growth rates: If individuals are classified by size or stage, we can estimate the probability that a surviving individual moves from its original class to each of the other potential classes.
- Fertility rates: The fertility rate is the average number of offspring that individuals in each class produce during the interval from one census to the next.

For the estimation of fertility rate the total number of viable seeds produced by each plant is needed. Flowering and fruiting patterns vary greatly from species to species and under different environmental conditions. If no previous information is available, it may be necessary to visit the population several times throughout the first year of monitoring to gather data on the phenology of the population. On these visits it is also important to note if the reproductive structures (flowers, inflorescences, fruits) remain on the plant once they mature or leave a trace that allows them to be counted at a later date. Information should also be gathered on whether flowering and fruiting take place gradually or synchronically within a plant and within the population. All these data will later be very helpful in estimating total fruit production per plant and total seed production. Normally total seed production per plant is estimated by multiplying total fruit production by the average number of viable seeds per fruit. When reproductive success depends strongly on the availability of suitable pollinators in the habitat, it may be necessary to include censuses of pollinators in the monitoring programme. Further information on seed dispersal, germination and dormancy may be needed to estimate the fertility rate from total seed production, although this information needs to be obtained only once or searched for in the literature and, therefore, is not to be included in the monitoring scheme.

The whole population or a representative sample of the population must be individually monitored to obtain data on vital rates. Individuals to be monitored are marked in a way that allows them to be re-identified at subsequent censuses. Aluminium or plastic tags can be attached to shrubs and trees with nails or wire while perennial herbaceous plants can be marked by inserting a tagged stake into the ground next to each individual. Tags should generally be inconspicuous to avoid attracting the attention of people and animals (Morris and Doak, 2002). Alternatively, specific detailed mapping procedures can be established that work best with herbaceous perennials. For instance, in a threatened population of *Erodium paularense* (see Box 4.1), a 1 × 0.5 m grid was laid on the monitoring plot. Subsequently, 1 × 0.5 m semi-rigid transparent PVC sheets were used in each cell to outline the cover of each *E. paularense* individual. The same PVC sheets were used in the following years to assess the survival and growth of each individual and the birth of new ones (Iriondo, 2003). At the time of marking or mapping, the state of each individual (stage, size, etc.) should also be recorded. The marked or mapped individuals must then be monitored at regular intervals. The flowering and fruiting periods are the best time to perform this task.

Table 4.1. Demographic parameters that can be used in a monitoring programme for genetic reserves in relation to population size and threat status of the target taxon.

Population size	Threat status	Density, frequency or cover	Population structure	Vital rates through individual monitoring
Large (>5000)	Non-threatened	×		
Small (<5000)	With no evident threats	×	×	
Small (<5000)	Threatened	×	×	×

Spatial structure

The spatial structure of a population provides information on the location of each individual. This information is very valuable for estimating the area of occupancy and assessing the relevance of within-population and interspecific competition and facilitation. Depending on the spatial scale of the plants, the coordinates of each individual can be gathered using standard topographic equipment, high-precision differential GPS or simply using a compass and a tape measure.

Selection and use of demographic parameters

The selection of which demographic parameters to use in monitoring will depend on the status of the target CWR populations, the species' life form and characteristics, and the planned data analysis. Tables 4.1 and 4.2 provide some general recommendations.

4.2.2 Ecological monitoring

Ecological monitoring identifies changes in the physical conditions that operate in the population and characterize the dynamics in the composition of communities associated with target CWR. When demographic data are not available, ecological monitoring may be a useful surrogate to infer population trends.

The microhabitat of plant populations is made up of many biotic and abiotic components and their importance varies in both space and time. Quantifying the effect of the environment on a plant requires measurement of both the plant and environmental factor of interest.

Abiotic components

Many abiotic factors can also be considered in a monitoring programme. Decisions about which components of the physical environment to measure will largely depend on the microhabitat of the CWR target population and the existing threats. Considering the time and expense involved, the monitoring of an abiotic factor should not be included in the monitoring programme unless there are sound reasons to justify this measure. A set of standard guidelines for measuring and reporting some of the most common components of the abiotic environment is provided in Gibson (2002).

Table 4.2. Use of density, frequency and cover according to the life form of the target taxon. (Adapted from Elzinga *et al.*, 1998.)

Life form	Frequency	Density	Cover
Annuals and biennials (e.g. *Avena, Daucus*)	Great changes can be expected from year to year. Frequency values will mainly be affected by changes in spatial distribution.	Permanent plots may be of little value if recruitment is spatially variable from year to year. Density values are prone to be affected by changes in environmental conditions.	Cover is affected by changes both in plant density and vigour. Cover values will be greatly affected by annual changes in environmental conditions.
Geophyte (e.g. *Narcissus*)	Changes in frequency values can mainly be attributed to changes in density. Dormancy and vegetative reproduction must be taken into account.	Density counts are often difficult due to vegetative reproduction. Flowering scapes may be a substitute for individuals.	Morphology is not suited for cover estimates. Changes in cover may be due to annual weather variation.
Herbaceous perennial, countable individuals (e.g. *Brassica oleracea*)	Changes in frequency may be caused by changes in density and/or spatial pattern. Long-lived species experience fewer spatial changes.	Measurements by class may be useful in interpretation.	Cover values will change more rapidly in short-lived species. Long-lived species are less sensitive to annual weather variation.
Herbaceous perennial, uncountable individuals (e.g. *Fragaria vesca*)	Changes in frequency may be caused by changes in density and/or spatial pattern. Long-lived species experience fewer spatial changes.	Density is difficult to use for this type of plant. It may be possible to use clumps as a counting unit.	Canopy cover is the most common measure for these plants.
Shrubs with multiple trunks (e.g. *Rubus idaeus*)	Rarely used except for seedlings. Plots would need to be large enough to achieve reasonable frequency.	Stems can be counted but it may be difficult to identify genets (individual plants). Changes in stems may be a more sensitive measure than changes in individuals.	Line intercept method is commonly used for shrubs.
Trees with isolated trunks (e.g. *Malus sylvestris*)	Frequency may be insensitive to changes that can be detected within the lifetime of the observer. Conversely, measurable changes are likely to be ecologically important.	Individuals are easy to count. Density may be insensitive to changes that can be detected within the lifetime of the observer. Conversely, measurable changes are likely to be ecologically important.	Line intercept method is commonly used for trees. Basal area based on the diameter at breast height is also common. Changes in cover that are measurable over a few years are likely to be important.

Climate is usually described in terms of the familiar elements of the weather. Temperatures and precipitation are the essential indicators, but others include solar radiation, wind, cloud cover, atmospheric pressure and humidity. Ideally, these indicators should be measured in the genetic reserve by establishing a weather station with the different sensors needed to register this information and a data logger to store the data. When this is not an option, climate data can be inferred from information gathered from the closest public meteorological stations. When these elements are measured systematically at a site over a period of several years, an accurate summary of the climate of the genetic reserve can be obtained. Using a variety of statistical techniques, we can compute averages for different climate elements as well as measures of variability and the frequency of more extreme events.

Abiotic components commonly measured in soil include moisture, texture, pH, nutrients, salinity, redox potential and cation exchange capacity. Good general descriptions of methodologies for describing and analysing the soil pertinent for plant studies are provided in Robertson *et al.* (1999), Pansu *et al.* (2001), Benton-Jones (2001) and Tan (2005).

Biotic components

The biotic environment comprises the living components of a plant's habitat. This ranges from neighbouring plants of the same species within a population to organisms at different trophic levels.

The plants that make up the community can establish competition and facilitate interactions with the target population. Plant community also influences physical and chemical factors in the habitat. Microclimate, light penetration and soil conditions are largely determined by the dominant plants, which also afford protection, and feeding and nesting sites for animals that may interact with the target populations. Plant community in the microhabitat is best described by obtaining density, cover and frequency values for each species as defined earlier. These values can then be expressed in an absolute or relative form. Relative values for density, dominance and frequency can be combined into a single *importance value*, which reflects these three somewhat different measures of species importance in the community.

Cover is normally recorded as an estimate of the percentage of the plot covered by each species. This is usually done by assigning cover-class estimates (Goldsmith *et al.*, 1986). Several scales for ranking cover have been suggested. One of the most common, which gives species with less cover proportionally greater weight than those with greater cover, is provided in Table 4.3.

Table 4.3. Cover classes and intervals. (From Cox, 1990.)

Cover class	Range of percent cover
1	0–1%
2	1–5%
3	5–25%
4	25–50%
5	50–75%
6	75–100%

Box 4.2. Parameters and equations for monitoring plant community structure. (From Cox, 1990.)

Density = no. of individuals / area sampled
Relative density = (density for a species / total density for all species) × 100
Dominance = total of basal area or aerial coverage values / area sampled
Relative dominance = (dominance for a species / total dominance for all species) × 100
Frequency = number of plots in which a species occurs / total number of plots sampled
Relative frequency = (frequency value for a species / total frequency values for all species) × 100
Importance value = relative density + relative dominance + relative frequency

Box 4.2 provides some of the most typical parameters used for monitoring plant community structure, while Table 4.4 presents a sample data sheet for monitoring plant community structure. The detection of significant changes in any of these variables through time may be a clear indication of changes in the habitat that may be affecting the viability of the target population.

Among mutualists, pollinators and seed dispersers are essential components of the reproductive process of many plants. Reproductive success of the target CWR species may be greatly dependent on the availability of these plant mutualists. When there are indications of limitation in pollinator and seed-disperser services, it may be useful to include periodical inventories and censuses of these animals in the monitoring programme. They will basically include information on density and frequency. Further information on these techniques can be found in Dafni *et al.* (2005). Dot blot or enzyme-linked immunosorbent assay (ELISA) techniques are used for determining the abundance of fungal mycorrhizae and *Rhizobium* bacteria (Perotto *et al.*, 1994).

When plant predators and parasites are suspected to significantly affect the vital rates of the target CWR population, similar approaches can be followed, gathering data on density and frequency for each species. Specific census techniques are available for birds, large and small mammals, insects, etc. (Elzinga *et al.*, 2001).

Table 4.4. Sample data sheet for monitoring plant community structure. Cover class (see Table 4.3). Aggregation class: 1: isolated individuals, 2: individuals in small groups, 3: individuals in relatively large groups, 4: lax somewhat continuous populations, 5: highly aggregated continuous populations.

Taxon name	Cover class	Aggregation class	Herbarium specimen code

The presence of pathogens in plants can often be determined by standard visual observations. Symptoms include changes in leaf colour, mottling, burns, stem and shoot damage, etc. There are a number of general texts and guidebooks available for visual identification of pathogens (e.g. Gram and Bovien, 1969; Fahy and Persely, 1983) and Internet resources with a general scope or related to specific crops (e.g. http://nu-distance. unl.edu/homer/public.html, http://plantpathology. tamu. edu/Texlab/index.htm). If the population size of a target CWR species is declining due to a pathogen, it may be advisable to determine the presence and abundance of the pathogen in the plants' tissues. There are accurate methods for pathogen detection such as histochemical, immunological and molecular methods. However, as all of these techniques require special equipment and knowledge, a laboratory at a university, agronomic public service or private company should be contacted to carry out these analyses. General texts providing detailed sampling methods for pathogens include Hampton *et al.* (1990), Dhingra and Sinclair (1995) and Narayanasamy (2001).

The intensity of pathogen infection or insect attached can be monitored using a simple ranking scale based on, for example, the proportion of the individuals in the population affected and the severity of the attack among a subsample of infected plants.

Disturbance and control sites

In addition to the above-mentioned parameters, ecological monitoring should report any natural or human-induced disturbance. Natural disturbance may include fire, flooding, slope movement, wind damage, extreme temperatures, trampling by animals and erosion. Examples of human causes of disturbance are mining, logging, domestic livestock grazing, recreation, road construction or maintenance and weed control. A standardized list of human causes of disturbance is included in the Unified Classification of Stresses by IUCN available at http://conservationmeasures.org/CMP/Site_Docs/IUCN- CMP_Unified_Direct_Threats_Classification_2006_06_01.pdf.

To account for changes in habitats and populations caused by factors that are beyond human control, it is also necessary to monitor sites where there are no active species-specific management interventions. Commonly referred to as control sites, they provide a baseline against which changes can be measured.

Climate change

CWR monitoring in genetic reserves must be capable of detecting and predicting changes in species composition and plant migrations caused by climate change, habitat destruction and altered disturbance regimes. Monitoring systems should also be able to assess efforts to minimize the impacts of climate change. Species-level impacts can be evaluated by tracking the distributional ranges of species over time as well as the timing of seasonal cycles and population growth rates. Along with information on local climate, these data can provide evidence that climate change is affecting species distributions or viability. In some cases, correlational studies may need to be supported by experimental studies.

The monitoring process involves selecting indicators such as susceptible species and habitats, and instituting annual recordings of the locations and timing of key events. Species to be used as indicators of climate change include those with distributions, physiology or life cycles that are sensitive to climate (especially temperature and precipitation) but less vulnerable to other environmental changes such as land-use shifts and pollution. Climate-related changes will be detected earlier at the boundary of a species' geographic range than at its centre. Shifts in the boundaries between vegetation types (ecotones) may be particularly sensitive indicators.

Phenology is perhaps the simplest approach to track species' responses to climate change because many processes are triggered by temperature-related cues (Root *et al.*, 2003). Common changes include earlier shooting and flowering, fruit ripening, hibernation, breeding and migration. Species that reproduce several times a year, such as some weedy CWR, may be able to produce more generations per year. Data on phenology can complement data on species abundance and distribution because they allow predictions to be made based on assumptions about the nature and rate of climate change. In addition, phenological indicators can also predict the functioning of communities and ecosystems because climate change can disrupt biological interactions.

Changing climate conditions can also cause changes in species composition through the replacement of a dominant species by a more tolerant subdominant species (*in situ* conversion). Changes in composition can also occur through migration of species from other areas. Migrating species may outcompete more sedentary, late successional or endemic species, increasing the risk of local extinction. In transitional zones, significant species mixing is likely (Thuiller *et al.*, 2005).

4.2.3 Anthropogenic information

The conservation of plant genetic diversity, including CWR, takes place within landscapes that are influenced by people in different ways and to different extents. Therefore, monitoring the biological status and effectiveness of conservation actions should incorporate social, economic, political and cultural threats and opportunities (Stem *et al.*, 2005). It is also important to engage stakeholders in conservation management and monitoring. Adjacent-dwelling communities that have traditionally used the goods and services found in genetic reserves often have first-hand knowledge and understanding of the status and behaviours of these areas' ecosystems and species. Engaging these communities in monitoring will allow more frequent and cost-effective collection of information.

4.2.4 Timing and frequency

Accurate monitoring results are also dependent on the timing and frequency of monitoring. Monitoring is most effective when it is timed to the seasons in which the target species is easiest to locate. For many species, the optimal time is when the species is in flower. For others, the fruiting stage might be most conspicuous. In some species, leaves are unusually coloured when young or when they are about to fall. For rare perennial spring wild flowers, monitoring may be best done early in the season before other vegetation grows and obscures the target species (Primack, 1998).

Monitoring that uses classes such as seedlings or reproductive individuals, or that focuses on certain size classes, should be scheduled so that annual measurements are done at the same phenological time each year. Otherwise, the number of individuals included in each class will usually change as the season progresses.

How often a target population should be monitored depends on the life form of the species and the expected rate of change, the rarity and trend of the species, and the resources available for monitoring. However, the main determinant of the frequency of sampling may be the strength and nature of the perceived threat to the population (Maxted *et al.*, 1997). For rare or very threatened annuals, monitoring may occur as often as every week or fortnight during each of several growing seasons, in particular if the demography of the population needs to be studied in detail (Harper, 1977). For perennial species, the interval between observations of adult individuals may well be several years, though there may be frequent monitoring of seedlings and saplings during a growing season to assess recruitment (Maxted *et al.*, 1997).

In practice, the frequency of sampling is likely to be much higher in a newly established reserve than in a well-established one. As the most appropriate management prescription will probably not be known when the reserve is established, initial monitoring frequency is likely to be high. However, as changes or adjustments to the management prescription become less necessary, the interval between monitoring exercises may be extended (Maxted *et al.*, 1997).

4.2.5 Consistency

Data collection methods need to be consistent throughout each census or sampling of the target CWR population. If a team is carrying out the monitoring, each member of the team needs to be clearly instructed and trained on the methods to be followed. This is especially important in monitoring programmes including a network of protected areas. The same procedures must be followed at each site to ensure that results reflect the real status of the populations being monitored and to allow for comparisons. The methods must be carefully written out so that measurements taken in successive time steps are done in the exact same manner. A comparable level of effort in successive time steps is required so that, for instance, changes in population size are due to real changes in plant numbers rather than to the intensity of searching by the census team (Primack, 1998).

4.2.6 Data analysis

Once field surveying is completed, the data provided by the monitoring process need to be properly analysed to be able to reach meaningful conclusions. Statistical tests to analyse information will be chosen when the monitoring programme is designed, observing that assumptions for proper use of the tests are met. If data are collected in the field with the expectation that some type of analysis can be carried out later, there is a high risk that the collected data cannot answer the questions originally posed, wasting a lot of time, effort and money. To prevent this, a statistician should be consulted during the design phase if those developing the monitoring programme lack knowledge in statistics. Essentially, the pilot study for a monitoring programme is conducted to help ensure that meaningful statistical tests are feasible.

It is also important to analyse data after each monitoring cycle. Data should not be stored for several years before analysis. Timely analysis can identify problems early on and ensure that questions requiring additional field visits or a different methodology can be addressed. In addition, problems that arise when data are being processed can often be answered when the fieldwork is still fresh in the data collector's memory.

One of the most essential data analyses for genetic reserve management is target CWR population viability analysis (PVA). Therefore, we are now going to describe it in further detail.

Population viability analysis

PVA is the use of demographic modelling methods to predict the future status of a population and help make conservation and management decisions. Viability is a probabilistic concept and, therefore, the future status of a population essentially refers to the likelihood that the population will be above some minimum size at a given future time.

Having conducted a demographic study, divided the population into classes and estimated class-specific vital rates to reflect the fate of individuals in all stages of the plant life cycle, we can construct a model to project the population into the future (see Box 4.1). Based on this model, the viability of the population can be assessed. A key element in population modelling and viability analysis of plant populations is the transition matrix. Each cell in the transition matrix table represents the probability that an individual will move to another class (Table 4.5 and Fig. 4.2). The class structure of a population at a given time can also be represented as a column matrix. Thus, the population structure at the next time step can be obtained by multiplying the transition matrix by the original population structure matrix. Subsequently, each successive year can be projected by replacing the previous population values in the population structure matrix with the most recently calculated values. In this manner, populations can be projected many years into the future.

However, with only a single transition matrix, the effect of environmental stochasticity on population viability cannot be assessed, so ideally demographic

Table 4.5. Transition matrix of a subpopulation of *Erodium paularense* growing on lithosol microhabitat. The subpopulation has been structured in four classes. The matrix contains the survival, growth and fertility rates corresponding to the life cycle in Fig. 4.2. (Adapted from Albert *et al.*, 2004.)

	Juvenile	Adult I	Adult II	Adult III
Juvenile	0.28	0.10	0.10	0.18
Adult I	0.28	0.68	0.12	0
Adult II	0	0.19	0.76	0.19
Adult III	0	0	0.10	0.80

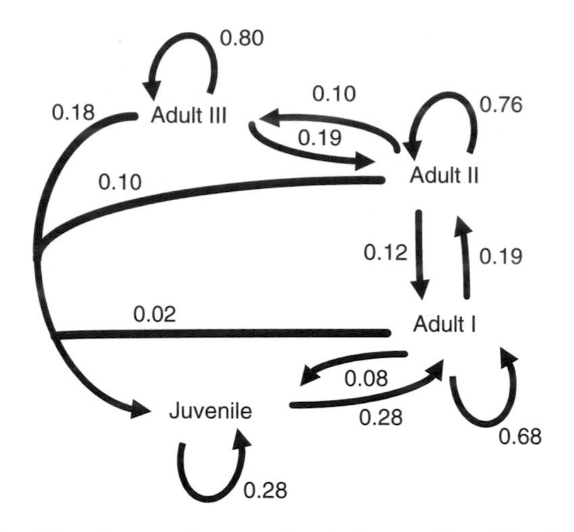

Fig. 4.2. Life cycle of *Erodium paularense*. Survival, growth and fertility rates in lithosol microhabitat.

data should be collected for several years. The term environmental stochasticity describes temporal variation in vital rates driven by unpredictable changes in the environment, such as changes in rainfall, temperature, pollinator, predator or pathogen density or quality.

Projection matrix models provide estimates of population growth and extinction risk. These data do not explicitly account for the underlying causes of population growth. Nevertheless, they provide information on whether a population is growing or declining in most years and, thus, whether further research is needed to understand and reverse any downward trends (Morris and Doak, 2002). One of the most important parameters that can be calculated for conservation purposes is lambda (λ), also called the finite rate of increase. Lambda values greater than one correspond to populations that are increasing, values equal to one represent populations that are stable, whereas values below one correspond to populations experiencing decline. Although estimates of mean performance of parameters such as lambda are the centrepiece of any PVA, the variability around mean rates is also key to determining viability, as greater variability involves higher risk of extinction.

If multiple CWR populations are being considered, spatial variability in vital rates and population growth must also be taken into account. The most plausible situation when dealing with multiple sites is that the mean and variance of vital rates, and hence of population growth rates, will differ across sites. Spatial differences are handled in a straightforward manner by estimating the means and variances of vital rates individually for each site (Morris and Doak, 2002).

One key issue in the estimation of extinction risks is the setting of quasi-extinction thresholds. PVA should not only consider the risk of population extinction, but should also assess the probability of population size falling below a particular threshold (Table 4.6). This is because many ecological and genetic processes affect the behaviour of populations when their size becomes very small. Therefore, an

Table 4.6. Effect of potential future threats and population management actions on quasi-extinction risk, expected abundance and probability of a 10% decline in population abundance: quasi-extinction risk was calculated considering a threshold of 500 individuals within a 30-year interval; abundance for which there is a 50% probability that the population will fall below this value at least once within a 30-year interval; and probability of a 10% decline in population abundance within a 10-year interval. (Adapted from Albert *et al.*, 2004.)

Scenario	Quasi-extinction risk (%)	Expected abundance (no. of plants)	Probability of a 10% decline
Present situation with basic PVA model (control)	57.0	493.1	100
Increase in environmental variability	80.8	444.3	99.8
Catastrophic regional droughts	88.3	305.2	100
Preventing human and cattle access, and reinforcement through sowing	0	1078.6	0
Preventing human and cattle access, and reinforcement through planting	0	965.6	9

important goal of PVA is to predict quasi-extinction, i.e. the probability of reaching a population size below which the population is likely to be critically and immediately imperilled (Ginzberg *et al.*, 1982).

Management options can be assessed in PVA through sensitivity analyses, where we evaluate how sensitive population growth is to particular demographic changes (Table 4.6). Thus, transition values within the matrix can be replaced with other hypothetical values to determine their effect on population projection. For example, what happens to the population growth rate (λ) of *Vicia bifoliolata* if seedling survival is increased by eliminating invasive *Carpobrotus* sp. plants that grow in the same microhabitat? Simulation of population response to changes in the transition values can identify areas where management can be most effective. Thus, sensitivity analysis can suggest different management options and evaluate the effectiveness of these alternatives. In essence, any management for improved population viability is an effort to change the mean and variance of population growth. Specific interventions, such as the control of competitive invasive species, almost always target a subset of the life stages of a species and even particular vital rates for that subset (e.g. survival of adults as opposed to their reproduction), hoping to influence overall population performance by changing these particular parts of the life cycle. An important consideration is that management interventions can have both positive and negative effects on population viability through effects on different vital rates (e.g. eliminating *Carpobrotus* sp. may promote recruitment of *V. bifoliolata* seedlings, but may increase herbivory pressure on adults). Since a transition matrix integrates vital rates to yield an estimate of population growth, it provides a natural way of determining how effective different management efforts are likely to be in changing population growth, before going to the time and expense of initiating them. This kind of analysis can also be used to assess the importance of uncertainty in the estimates of different vital rates for predictions about population viability. If population growth is highly influenced by the exact value of adult survival, having an accurate estimate of this rate is key for proper viability assessment (Morris and Doak, 2002).

Another useful parameter that can be calculated from a transition matrix is elasticity. Elasticity values are a measure of the sensitivity of the population growth rate to a relative change in the transition probability (de Kroon *et al.*, 1986; Caswell, 2001). For an equal relative change in transition probability, transitions with high elasticity values cause a greater change in the population growth rate than those with a low elasticity value. Elasticity values can identify which stages and transitions should be managed to provide the largest overall population benefits. However, the effect of a particular transition on population growth also depends on the natural variability of its probability value. Thus, for some species, regardless of elasticity values, the transitions with the highest natural variability may be the most limiting to population growth or survival (Schemske *et al.*, 1994).

Practical reviews of these methods applied to specific plant study cases can be found in many books and journals (e.g. Picó, 2002; Iriondo, 2003; Tremblay and Hutchings, 2003; Albert *et al.*, 2004). Several software packages are also available for this type of demographic studies, such as RAMAS (http://www.ramas.com/), VORTEX (http://www.vortex9.org/vortex.html) and POPTOOLS (http://www.cse.csiro.au/poptools/).

4.3 Genetic Monitoring Methodologies

4.3.1 Background to genetic reserve monitoring

The aim of plant genetic conservation is often stated as being to conserve, as far as possible, the range of genetic diversity found in a target species. This aim recognizes that genetic diversity is a critical component of biodiversity per se and that the genetic resources themselves are a rich potential source of useful genetic traits. However, in addition to the preservation of genetic material, genetic conservation needs to maintain the evolutionary process. This is best achieved through *in situ* conservation where selective forces and adaptive genetic change will be allowed to operate. Whether either or both of these objectives are being targeted, it is important that genetic diversity is recognized, studied and measured in certain circumstances at least.

Woodruff (1992) rightly stated that ecological management is the most pragmatic (cheapest and most effective) way of conserving genetic diversity, indicating that:

> genetic factors do not figure among the four major causes of extinction (Diamond, 1989). Thus, although genetic factors are major determinants of a population's long-term viability, conservationists can do more for a threatened population in the short term by managing its ecology.

Therefore, it is neither possible nor in many cases desirable to undertake genetic monitoring of CWR populations that are conserved *in situ*. Within Europe we have as many as 20,000 CWR taxa, causes of genetic erosion of their diversity are many and in reality we have very little information on their population distributions and frequencies, let alone genetic information. Resources, both human and financial, cannot and will not allow the monitoring of large numbers of populations for all or even most of our CWR taxa. Genetically-based approaches to conservation monitoring can, therefore, only be applied to the most highly prioritized taxa, and only after proxy or surrogate measures of genetic diversity have been applied. Earlier in this chapter population monitoring methods have been discussed, and in many cases populations that are managed in different geographical locations under different environmental conditions will be adequate surrogates for genetic diversity of a species. Equally, management regimes that aim to maintain as many infraspecific taxa of a target species as possible will clearly maintain by proxy genetic diversity existing below the species level but above the population level. Further, where taxonomically complex groups (TCGs) (a CWR example would be *Sorbus* spp.) present difficulties in some respects, they can actually be welcomed as a proxy means of conserving genetic diversity, provided that the genetic reserve management guidelines are aimed at conserving evolutionary processes (Ennos *et al.*, 2005) that will allow for the dynamic flux of the subgroups that may exist within a TCG. It may well be possible to conserve the components of a TCG together with the necessary evolutionary processes without resorting to complex molecular and population genetic studies.

Sophisticated molecular genetic techniques will, therefore, only be used for 'fine-tuning' the development of our priorities for conservation and for making assessments of populations in extreme situations of threat.

4.3.2 Key population genetic issues and parameters

Fitness

Reproductive fitness is the measure of an individual's ability to contribute offspring or progeny to the subsequent generation. For sexually reproducing species, fitness traits will encompass an individual plant's ability to participate in the sexual cycle in terms of production of male and female gametes, fertile seeds and the number of reproducing plants produced from those seeds. Traits contributing to fitness are generally quantitative traits and are, therefore, difficult to measure. Whether or not we can easily measure them, for plant species that are generally cross-pollinating, they are likely to be subject to inbreeding depression if inbreeding reduces the frequency of heterozygotes in the population, something which is measurable in population genetic terms (see below). Significant reductions in fitness traits associated with viability or fecundity are obviously undesirable, and so a minimum population size needs to be maintained to reduce inbreeding depression to an acceptably low level. Maintenance of heterozygosity in a population which is not necessarily of adequate minimum size, and also its fitness, can be achieved by gene flow from other populations (the level estimated, e.g. by *F* statistics), provided that the 'genetic diversity' that is introduced in this way does not contain maladaptive genes.

The other factor to take into account is that the expression of fitness traits will be dependent upon the environment (and fitness traits generally have low heritabilities), so change in environmental conditions will obviously affect fitness, but not in an easily measurable way. Environmental change may gradually eliminate a population, or the population may be able to maintain its fitness by adapting to the change, but this will depend upon adequate levels of genetic variation existing in the population.

Effective population size

This can broadly be defined as 'the size of an idealized (hypothetical) population that would lose genetic diversity at the same rate as the actual population (the one under study)'. Many or most plants set at least some of their seeds by self-pollination. Populations of plants often vary in size from one generation to the next. The number of genes that individual plants contribute to the next generation is rarely random, some producing much more seed than others. All these factors make populations genetically smaller than their actual or census size. This smaller size is called the effective population size, *Ne*, and it is more important than *N*, the actual size, when considering conservation (Wright, 1931). Unfortunately, estimating effective population size is extremely difficult because of obtaining reliable estimates of the factors mentioned above, and the way that they interact. *Ne* will be closest to *N* in very large outbreeding populations where all plants produce roughly the same number of seeds, and where the number of plants remains more or less constant from one generation to the next. However, *Ne* will be much smaller than *N* for species that are substantially inbreeding, where a few plants produce most of the seed or where a population suffers a drastic, albeit temporary, reduction in numbers. So, just taking the example of a predominantly self-pollinating species, *Ne* could be close to half of *N*. What happens if *N* varies over generations? Suppose in five successive generations a population has 50, 50, 10, 50 and 50 plants in it, then *Ne* is actually as low as 27! (see Frankham *et al.*, 2004). As a rule of thumb, *Ne* is controlled by the lowest number (as a harmonic mean); when *N* varies among generations, a slight increase in the lowest generation has an impact. Similarly, for a dioecious species, *Ne* is determined by the rarest sex in the population.

In summary, effective population size in relation to actual population size will always be smaller for inbreeding rather than outbreeding species, smaller if just a few plants contribute most to the next generation and smaller if the actual population size drops substantially even for just one generation.

Genetic diversity, gene flow and population structure

The concept of 'diversity' may seem simple, but it is not; this can lead to the unnecessary and inappropriate use of some measures of diversity, of which there are two possibilities: the first is 'richness' – the total number of genotypes or alleles present within germplasm regardless of frequency; and the second is 'evenness' of the frequencies of different alleles or genotypes. Where richness is used to measure diversity, germplasm with more (and different) alleles or genotypes will be more diverse. In contrast, where evenness is considered important, a germplasm sample where the alleles or genotypes are all roughly equal in frequency will be more diverse than one where there are the same numbers of alleles or genotypes, but where they are very unequal in frequency.

Estimations of allelic richness are, therefore, the number of distinct alleles at a locus (A), and estimations of diversity (H) are measures based upon the frequencies of variants: allelic variants in the case of Nei's index of diversity (Nei, 1973) and phenotypic variants in the case of the Shannon–Weaver index (H). Such estimators can be used to assess diversity occurring within populations, within reserves or protected areas, or within different geographical regions. Thus, genetic diversity measures enable comparisons of genetic diversity in candidate populations to be made in readiness for setting up a genetic reserve, or for monitoring what happens to diversity in one reserve over a particular time period.

It is necessary to make reference to both 'observed' and 'expected' heterozygosities. The Hardy–Weinberg expected heterozygosity (for a locus with two alleles is also called gene diversity) can be calculated to measure genetic diversity. For conservation purposes, genetic diversity will be averaged over several or many loci, which will be the average heterozygosity. Here, reference is made to the Hardy–Weinberg equilibrium, the detail of which can be found in most textbooks (see, e.g. Frankham *et al.*, 2004). In essence, this describes the situation for an entirely hypothetical population which is extremely large, which is undergoing random mating and where there is no mutation, selection or migration. Under these conditions, allele and genotype frequencies are in equilibrium after one generation. What is important about this for conservation is that it provides a baseline or a means by which levels of migration or selection, levels of non-random mating or reduced effective population sizes can be detected and estimated in real populations. So, deviations of the observed genotype frequencies from the Hardy–Weinberg expected genotype frequencies can indicate that there are specific factors affecting the genetic structure of the population.

As will be seen below, the estimation of gene flow between populations or subpopulations, inbreeding within populations or subpopulations and differentiation between subpopulations can be important and informative in certain situations for CWR management. This is most commonly achieved using what are called F statistics which are related to observed and average heterozygosity (see above). F_{ST} measures inbreeding due to differentiation among subpopulations relative to the total population, F_{IS} measures the inbreeding of individuals relative to the subpopulation and F_{IT} measures the inbreeding of the total population. Here we are considering inbreeding in a broad sense, as the mating of individuals related by ancestry, measured by the probability that two alleles at a locus are identical by descent. There are many examples of the uses of these estimators and many useful references, but a good starting point would be Frankham *et al.* (2004). Questions that can be answered may be:

- Are two populations or subpopulations sufficiently differentiated for both to receive conservation attention?

- Is the level of inbreeding within a conserved population increasing (leading to the questions why, and what can be done about it)?
- Is a particular species characterized by having high levels of inbreeding?

How can such analyses be undertaken? There are now many examples of software that can be downloaded free from the web which will calculate the parameters listed above. Examples of such packages are ARLEQUIN, GDA (Genetic Data Analysis), FSTAT, GENEPOP, POPGENE and TFPGA (Tools for Population Genetic Analysis). There are also many web sites that will link you to these software packages and more. One such example is available at http://www.biology.lsu.edu/general/software.html. These packages differ somewhat in the estimated parameters and computational approaches, but most are effective. One of the differences between these packages is the data format required, and the ease with which data sets can be prepared for analysis. Some, however, conform to the NEXUS data format, and this is clearly an advantage (for a very useful review see Excoffier and Heckel, 2006).

Minimum viable population

What is the minimum size of a population needed to remain genetically viable and maintain genetic variation and heterozygosity? This minimum size, allowing for substantial variation in definition, is commonly referred to as the MVP size. It is generally accepted, first, that the minimum size of a population in which inbreeding depression is reduced to an acceptably low level is an MVP size of 50 individuals. Second, an MVP of 500 individuals should be of sufficient size to allow new variation arising from mutation to replace that lost by genetic drift (Franklin, 1980). This is the '50/500 rule'. When dealing with natural populations, the 50/500 rule concerns *Ne*, the effective population size, rather than *N*, the actual population size. Allowing for this, the 50/500 rule then becomes the 500/5000 rule. In short, an MVP size of 5000 is probably reasonably safe, but is tentative and could be reduced if better estimates of the *Ne*/*N* ratio were available for the species and showed that *Ne* was indeed close to *N*. Analysis of the factors that make *Ne*/*N* less than one shows that fluctuation in population size is the most important.

4.3.3 Overview and choice of molecular markers for genetic monitoring

The range of molecular markers that can now be used relatively easily is quite extensive. First, DNA-based markers have substantially overtaken those based upon proteins. Techniques applied to studying plant populations can include identifying polymorphisms in the actual DNA sequence, identifying restriction fragment length polymorphisms (RFLPs), and the use of polymerase chain reaction (PCR)-based technology to find polymorphisms using random amplified polymorphic DNA (RAPD), simple sequence repeat (SSR) polymorphism, or combination techniques such as amplified fragment length polymorphism (AFLP). While reviews of these techniques are plentiful (Karp *et al.*, 1997; Newbury and Ford-Lloyd, 1997; Westman and Kresovich, 1997; Nybom, 2004), because of the rapidity with which relevant technology is proceeding, they may not remain comprehensive for long. For example, single nucleotide polymorphisms (SNPs) or microarrays that can generate e-markers (expression markers) are becoming more widely used.

Nevertheless, it is possible to identify important criteria by which to judge the value of any particular marker system to a chosen application. For instance, for information on population history or phylogenetic relationships, sequence data may be most appropriate. From how much DNA sequence or from how many loci will information be required, and how much allelic variation per locus is appropriate? To what extent are the methodologies robust and reproducible? Is cost per data point an important factor, and how important is the speed of analysis? How much DNA is available (most methods involving PCR require only very small quantities of easily prepared DNA, while certain RFLP analyses require larger amounts)? Is it necessary to identify homozygotes and heterozygotes, in which case co-dominant markers are needed, or will dominant markers suffice? All of these questions may need to be answered before any study of germplasm is undertaken.

Desirable properties of molecular markers

- **Polymorphic:** The marker must be polymorphic, as the polymorphism itself is measured for genetic diversity studies. However, the level of polymorphism detected can vary depending on the type of molecular marker used.
- **Co-dominant inheritance:** The different allelic forms of the marker should be detectable in diploid organisms to allow the discrimination of homozygotes and heterozygotes.
- **Representative of all parts of the genome:** The markers should be evenly and frequently distributed throughout the genome for general studies of genetic diversity. However, it may be necessary to use markers which saturate some parts of the genome, or certain chromosomes (e.g. for fine genetic mapping). Alternatively, it may be appropriate to focus markers in genomic regions with lots of genes, or within expressed gene sequences (ESTs).
- **Easy, fast and cheap:** The markers should be easy, fast and cheap to detect.
- **Reproducible:** The markers should be reproducible within and between laboratories.

There is no single molecular marker system that meets all these needs, but for most population genetic studies co-dominant markers are clearly preferable because they allow the determination of the different alleles at a locus in heterozygotes and, hence, a precise determination of allele frequency. The markers that are appropriate and very frequently used are microsatellites, otherwise known as simple sequence repeats (SSRs) or simple tandem repeats (STRs).

Microsatellites are nucleotides that are repeated in tandem, where the repeat length can be between 1 and 6 nucleotides and which are found in all higher organisms. While the use of SSRs for studying diversity is relatively simple and based upon PCR technology, it is first necessary to detect the SSRs of a particular species and obtain sequence information on the flanking areas in order to design the primer pairs needed for the PCR reactions, and this is more complex. Data mining of DNA sequences in major database repositories of sequence data can be used to locate SSRs, but detection of microsatellites otherwise is commonly conducted by enriching genomic DNA libraries for particular repeat sequences followed by the sequencing of large numbers of clones. Out of 160 taxa randomly sampled from the Crop Wild Relative (CWR) Catalogue, only 29% were found to have published microsatellite primer sequences available. So for population monitoring of a majority of CWR taxa, primer sequences would need to be developed. As this is a commercially available service nowadays, technical know-how is not necessarily required – just the necessary funding.

While the one major disadvantage to using SSRs is the difficulty in identifying and isolating repeat sequences, the advantages are many:

- Easy to use once isolated and primers designed;
- Only needs a small amount of DNA for fingerprinting;
- Highly polymorphic;
- Co-dominant;
- Scattered throughout the genome;
- Can be focused on expressed genes if necessary;
- Can be reliably exchanged between laboratories – assays are robust and repeatable;
- Useful for high-throughput automation.

4.3.4 Decision making – how, when and why to use genetic monitoring?

How and when to use?

For molecular population genetic studies, random samples of leaves from individual plants can be taken. Sampling 50 plants is generally considered to be adequate, but useful results can be obtained by sampling as few as 20. Clearly, if any population substructure is known to exist (i.e. subpopulations), this would need to be taken into account, with 20–50 plants being sampled from each. When and how frequently to sample is entirely dependent upon the question that needs to be answered, and should be obvious in relation to the questions about 'why' below. However, the important message would be not to re-sample just for the sake of it, but to do so if there is good reason to think that a population was changing in some way as a result of some change in the environment.

Why to use?

If population monitoring is being undertaken for various reasons such as during the management of an existing protected area, or in order to determine what populations should receive protection, when and why is it worthwhile initiating a molecular population genetic study?

- To recognize situations where an overall reduction of fitness of a population might occur. Where levels of gene diversity in candidate populations can be compared using molecular markers before or during reserve design, those populations that are more susceptible to future reductions in fitness can be identified because they may have reduced levels of genetic diversity compared to others.

- As a pilot study ahead of protection in a genetic reserve, to determine the extent to which a species is inbreeding or outbreeding (*F* statistics – coefficient of inbreeding). If, in a species which is entirely outbreeding, variation is partitioned mostly within populations and very little between, there is least concern about which population(s) to choose for protection – most populations may well possess as much as 80% of the total variation. However, whether a species is predominantly inbreeding or outbreeding may already be known.
- To determine which inbreeding populations should be chosen for protection – which ones possess the most genetic variation (e.g. in the sense of Nei's diversity)? This should be used with caution as cost-effectiveness might not be high. Easier and cheaper alternatives will be to undertake an ecogeographic study to identify those populations which are most likely to contrast in terms of ecogeographic adaptation and, hence, be most genetically diverse. Also, those populations which represent different infraspecific taxa or which are identifiable as being morphologically different and a result possibly of microspeciation could be chosen for protection without the need to resort to molecular population genetics.
- When there must be, for whatever reason, a choice between which small isolated populations (whether inbreeding or outbreeding) should be protected, there is a very clear need to use molecular population genetics to help in the decision making. Important information that cannot be obtained by other proxy methods will include the observed and expected heterozygosity, the genetic diversity and the amount of gene flow occurring between the target and other populations.
- To determine what to do if a population being protected or being considered for protection has been subjected to a severe decline in population size. Population genetics already tells us that this is not good, and should be avoided because even if the population size increases again, there will have been a potentially damaging reduction in effective population size. However, we will not actually know what the effective population size is before or after the fall in size. Molecular population genetics could tell us whether the population has suffered a decline in genetic diversity or a decrease in observed heterozygosity (provided that we have something to compare with, i.e. an initial assessment of genetic diversity or a larger population elsewhere). In this case there is no alternative to molecular population genetic analysis, as we have no easy means of measuring effective population size accurately.
- How often should we undertake our regular population monitoring? It is almost certainly unnecessary to carry out routine monitoring, but it may be worthwhile undertaking an initial population genetic analysis of a sample of plants from the population(s) to be conserved. Subsequent analyses may then only be necessary if there is an observed fall in the actual population size, which can of course be determined by population monitoring rather than molecular population genetic monitoring.
- If populations are fragmented or become fragmented within a protected area or reserve, it is possible to establish the extent to which gene flow (*F* statistics)

occurs between the fragments and, therefore, how vulnerable they may be to loss of fitness. However, effective population monitoring will tell us whether populations are fragmented and potentially have an effective population size below that of the MVP.

In summary:

- **Do not** plan to do molecular population genetic monitoring first in any *in situ* conservation assessment.
- **Do not** undertake molecular population genetic assessment/monitoring without very good reason, or without specific questions to answer, and until other proxy genetic assessments have been fully examined.
- **Do not** necessarily plan for routine sequential molecular population genetic monitoring.
- **Do** use molecular population genetic assessment as a last resort and for fine-tuning to:
  - Select the most suitable and fittest populations for *in situ* conservation;
  - Measure inbreeding/outbreeding in a species as a pilot survey;
  - Monitor populations or critical situations;
  - Select for conservation among candidate populations of inbreeding species;
  - Select the 'best' small isolated populations for protection;
  - Determine the effect of a severe drop in actual population size on genetic diversity;
  - Establish whether gene flow is occurring between fragmented populations.

References

Albert, M.J., Escudero, A. and Iriondo, J.M. (2001) Female reproductive success of the narrow endemic *Erodium paularense* (Geraniaceae) in contrasting microhabitats. *Ecology* 82, 1734–1747.

Albert, M.J., Draper, D. and Iriondo, J.M. (2004) *Erodium paularense* in Spain: relevance of microhabitats in population dynamics. In: Akcakaya, H.R., Burgman, M., Kindvall, O., Wood, C.C., Sjogren-Gulve, P., Hartfield, J. and McCarthy, M. (eds) *Species Conservation and Management: Case Studies*. Oxford University Press, New York, pp. 75–89.

Benton-Jones, J. (2001) *Laboratory Guide for Conducting Soil Tests and Plant Analysis*. CRC Press, Boca Raton, Florida.

Bonham, C.D., Bousquin, S.G. and Tazik, D.J. (2001) *Protocols and Models for Inventory, Monitoring, and Management of Threatened and Endangered Plants*. Colorado State University, Fort Collins, Colorado.

Brady, W.W., Aldon, E.F. and Cook, J.W. (1993) *COVER: A Decision Support System for Monitoring Vegetation Cover Change on Forest Grasslands*. USDA Forest Service, Rocky Mountain Forest and Range Experiment Station, Fort Collins, Colorado.

Caswell, H. (2001) *Matrix Population Models: Construction, Analysis and Interpretation*, 2nd edn. Sinauer Associates, Sunderland, Massachusetts.

Charlesworth, D. and Pannell, J.R. (2001) Mating systems and population genetic structure in the light of coalescent theory. In: Silvertown, J. and Antonovics, J. (eds) *Integrating Ecology and Evolution in a Spatial Context*. Blackwell, London, pp. 73–95.

Cohen, J. (1988) *Statistical Power Analysis for the Behavioral Sciences*, 2nd edn. Lawrence Erlbaum, Hillsdale, New Jersey.

Cox, G.W. (1990) *Laboratory Manual of General Ecology*, 6th edn. Wm.C. Brown, Dubuque, Iowa.

Croy, C.D. and Dix, R.L. (1984) Note on sample size requirements in morphological plant ecology. *Ecology* 65, 662–666.

Dafni, A., Kevan, P.G. and Husband, B.C. (2005) *Practical Pollination Biology*. Enviroquest, Cambridge, Ontario, Canada.

de Kroon, H., Plaisier, A., van Groenendael, J. and Caswell, H. (1986) Elasticity: the relative contribution of demographic parameters to population growth rate. *Ecology* 67, 1427–1431.

DeVries, P.G. (1979) Line intersect sampling – statistical theory, applications, and suggestions for extended use in ecological inventory. In: Cormack, R.M., Patil, G.P. and Robson, D.S. (eds) *Sampling Biological Populations*, Vol. 5: *Statistical Ecology*. International Cooperative, Fairland, Maryland, pp. 1–70.

Dhingra, O.D. and Sinclair, J.B. (1995) *Basic Plant Pathology Methods*, 2nd edn. CRC Press, Boca Raton, Florida.

Diamond, J. (1989) Overview of recent extinctions. In: Western, D. and Pearl, M. (eds) *Conservation for the Twenty-first Century*. Oxford University Press, New York, pp. 37–41.

Dytham, C. (1999) *Choosing and Using Statistics: A Biologist's Guide*. Blackwell, Oxford.

Elzinga, C.L., Salzer, D.W. and Willoughby, J.W. (1998) *Measuring and Monitoring Plant Populations*. Bureau of Land Management, Denver, Colorado.

Elzinga, C.L., Salzer, D.W., Willoughby, J.W. and Gibbs, J.P. (2001) *Monitoring Plant and Animal Populations*. Blackwell, Malden, Massachusetts.

Ennos, R.A., French, G.C. and Hollingsworth, P.M. (2005) Conserving taxonomic complexity. *Trends in Ecology and Evolution* 20, 164–168.

Excoffier, L. and Heckel, G. (2006) Computer programs for population genetics data analysis: a survival guide. *Nature Reviews Genetics* 7, 745–758.

Fahy, P.C. and Persely, G.J. (1983) *Plant Bacterial Diseases: A Diagnostic Guide*. Academic Press, New York.

Floyd, D.A. and Anderson, J.E. (1987) A comparison of three methods for estimating plant cover. *Journal of Ecology* 75, 229–245.

Frankham, R., Balou, J.D. and Briscoe, D.A. (2004) *Introduction to Conservation Genetics*. Cambridge University Press, Cambridge.

Franklin, I.R. (1980) Evolutionary change in small populations. In: Soule, M.E. and Wilcox, B.A. (eds) *Conservation Biology: An Evolutionary–Ecological Perspective*. Sinauer Associates, Sunderland, Massachusetts, pp. 135–150.

Gibson, D.J. (2002) *Methods in Comparative Plant Population Ecology*. Oxford University Press, New York.

Ginzberg, L.R., Slobodkin, L.B., Johnson, K. and Bindman, A.G. (1982) Quasi-extinction probabilities as a measure of impact on population growth. *Risk Analysis* 2, 171–181.

Goldsmith, F.B., Harrison, C.M. and Morton, A.J. (1986) Description and analysis of vegetation. In: Moore, P.D. and Chapman, S.B. (eds) *Methods in Plant Ecology*. Blackwell, Oxford, pp. 437–524.

González-Benito, M.E., Martin, C. and Iriondo, J.M. (1995) Autecology and conservation of *Erodium paularense* Fdez. Glez. & Izco. *Biological Conservation* 72, 55–60.

Gram, E. and Bovien, P. (1969) *Recognition of Diseases and Pests of Farm Crops*. Blandford Press, London.

Hampton, R., Ball, E. and De Boer, S. (eds) (1990) Serological methods for detection and identification of viral and bacterial plant pathogens. *A Laboratory Manual*. American Phytopathological Society, St Paul, Minnesota.

Harper, J.L. (1977) *Population Biology of Plants*. Academic Press, London.

Hayek, L.C. and Buzas, M.A. (1997) *Surveying Natural Populations*. Columbia University Press, New York.

Iriondo, J.M. (2003) Aplicación de herramientas informáticas para la determinación de la viabilidad poblacional: Un caso práctico: estudio de las poblaciones de *Erodium paularense* en la Comunidad de Madrid. In: Bañares, A. (ed.) *Biología de la Conservación de Plantas Amenazadas*. Organismo Autónomo Parques Nacionales, Ministerio de Medio Ambiente, Madrid, Spain, pp. 163–175.

IUCN (2001). *IUCN Red List Categories and Criteria: Version 3.1*. IUCN Species Survival Commission. IUCN, Gland, Switzerland/Cambridge.

Karp, A., Kresovich, S., Bhat, K.V., Ayad, W.G. and Hodgkin, T. (1997) *Molecular Tools in Plant Genetic Resources Conservation: A Guide to the Technologies*. IPGRI Technical Bulletin 2, IPGRI, Rome, Italy.

Krebs, C.J. (1989) *Ecological Methodology*. Harper & Row, New York.

Lucas, H.A. and Seber, G.A.F. (1977) Estimating coverage and particle density using the line intercept method. *Biometricka* 64, 618–622.

Maxted, N., Guarino, L. and Dulloo, M.E. (1997) Management and monitoring. In: Maxted, N., Ford-Lloyd, B.V. and Hawkes, J.G. (eds) *Plant Genetic Conservation: The* In Situ *Approach*. Chapman & Hall, London, pp. 144–159.

McCall, C.H. Jr. (1982) *Sampling and Statistics Handbook for Research*. Iowa State University Press, Ames, Iowa.

Morris, W.F. and Doak, D.F. (2002) *Quantitative Conservation Biology: Theory and Practice of Population Viability Analysis*. Sinauer Associates, Sunderland, Massachusetts.

Narayanasamy, P. (2001) *Plant Pathogen Detection and Disease Diagnosis*. CRC Press, Boca Raton, Florida.

Nei, M. (1973) Analysis of gene diversity in subdivided populations. *Proceedings of the National Academy of Sciences USA* 70, 3321–3323.

Newbury, H.J. and Ford-Lloyd, B.V. (1997) Estimation of genetic diversity. In: Maxted, N., Ford-Lloyd, B.V. and Hawkes, J.G. (eds) *Plant Genetic Conservation: The* In Situ *Approach*. Chapman & Hall, London, pp. 192–206.

Nybom, H. (2004) Comparison of different nuclear DNA markers for estimating intraspecific genetic diversity in plants. *Molecular Ecology* 13, 1143–1155.

Pansu, M., Gautheyrou, J. and Loyer, J.Y. (2001) *Soil Analysis – Sampling, Instrumentation and Quality Control*. Balkema, Lisse, The Netherlands.

Perotto, S., Malavasi, F. and Butcher, G.W. (1994) Use of monoclonal antibodies to study mycorrhiza: present applications and perspectives. In: Norris, J.R., Read, D.J. and Varma, A.K. (eds) *Techniques for Mycorrhizal Research*. Academic Press, San Diego, California, pp. 681–708.

Picó, F.X. (2002) Desarrollo, análisis, e interpretación de los modelos demográficos matriciales para la biología de la conservación. *Ecosistemas* 11, 48–53.

Primack, R. (1998) Monitoring rare plants. *Plant Talk* 15, 29–32.

Ringold, P.L., Alegria, J., Czaplewski, R.L., Mulder, B.S., Tolle, T. and Burnett, K. (1996) Adaptive monitoring design for ecosystem management. *Ecological Applications* 6(3), 745–747.

Robertson, G.P., Coleman, D.C., Bledsoe, C.S. and Sollins, P. (1999) *Standard Soil Methods for Long-term Ecological Research*. Oxford University Press, New York.

Root, T.L., Price, J.T., Hall, K.R., Schneider, S.H., Rosenzweig, C. and Pounds, J.A. (2003) Fingerprints of global warming on wild animals and plants. *Nature* 42, 57–60.

Schemske, D.W., Husband, B.C., Ruckelshaus, M.H., Goodwillie, C., Parker, I.M. and Bishop, J.G. (1994) Evaluating approaches to the conservation of rare and endangered plants. *Ecology* 75, 584–606.

Silvertown, J.W. and Lovett-Doust, J. (1993) *Introduction to Plant Population Biology*. Blackwell, Oxford.

Stafford, S.G. (1993) Data everywhere but not a byte to read: managing monitoring information. *Environmental Monitoring and Assessment* 26, 125–141.

Steidl, R.J. and Thomas, L. (2001) Power analysis and experimental design. In: Scheiner, S.M. and Gurevitch, J. (eds) *Design and Analysis of Ecological Experiments*, 2nd edn. Oxford University Press, New York, pp. 14–36.

Stem, C., Margoluis, R., Salafsky, N. and Brown, M. (2005) Monitoring and evaluation in conservation: a review of trends and approaches. *Conservation Biology* 19(2), 295–309.

Tan, K.H. (2005) *Soil Sampling, Preparation, and Analysis*. CRC Press, Boca Raton, Florida.

Thomas, L. and Krebs, C.J. (1997) A review of statistical power analysis software. *Bulletin of the Ecological Society of America* 78, 128–139.

Thuiller, W., Lavorel, S., Araújo, M.B., Sykes, M.T. and Colin Prentice, I. (2005) Climate change threats to plant diversity in Europe. *Proceedings of the National Academy of Sciences USA* 102(23), 8245–8250.

Tremblay, R.L. and Hutchings, M.J. (2003) Population dynamics in orchid conservation: a review of analytical methods based on the rare species *Lepanthes eltoroensis*. In: Dixon, K.W., Kell, S., Barrett, R.L. and Cribb, P.J. (eds) *Orchid Conservation.* Natural History(Borneo), Kota Kinabalu, Sabah, Malaysia, pp. 183–204.

Westman, A.L. and Kresovich, S. (1997) Use of molecular marker techniques for description of plant genetic variation. In: Callow, J.A., Ford-Lloyd, B.V. and Newbury, H.J. (eds) *Biotechnology and Plant Genetic Resource.* CAB International, Wallingford, UK, pp. 9–48.

Woodruff, D. (1992) *Biodiversity Conservation and Genetics*. Proceedings of the 2nd Princess Chulabkorn Congress of Scientific Technology, Bangkok, Thailand.

Wright, S. (1931) Evolution in Mendelian populations. *Genetics* 16, 97–159.

5 Population and Habitat Recovery Techniques for the *In Situ* Conservation of Plant Genetic Diversity

S.P. Kell,[1] E. Laguna,[2] J.M. Iriondo[3] and M.E. Dulloo[4]

[1]*School of Biosciences, University of Birmingham, Edgbaston, Birmingham, UK;* [2]*Centro para la Investigación y Experimentación Forestal (CIEF), Generalitat Valenciana. Avda. País Valencià, Valencia, Spain;* [3]*Área de Biodiversidad y Conservación, Depto. Biología y Geología, ESCET, Universidad Rey Juan Carlos, Madrid, Spain;* [4]*Bioversity International, Rome, Italy*

5.1 Introduction

5.1.1 Recovery techniques for plant genetic diversity: establishing the context

The previous chapters have presented procedures and guidelines for the location and design of genetic reserves, approaches to management planning and implementation, and population monitoring for the *in situ* conservation of plant genetic diversity in protected areas. This chapter looks at the role of population and habitat recovery techniques. The main purposes are to outline the circumstances in which population and/or habitat recovery may be needed for the *in situ* conservation of plant genetic diversity (with an emphasis on crop wild relatives (CWR) in genetic reserves), to review the techniques available and to present the reader with a guide to accessing existing information sources that can be put to practical use during the implementation of the protected area and/or genetic reserve management plan.

The theory and practice of population and habitat recovery is often viewed as a specialist area that does not traditionally play a role in the conservation of plant genetic resources for food and agriculture (PGRFA) – the plants that are used in agriculture and which contribute to agricultural sustainability and crop improvement, including CWR. There is a plethora of papers, books, organizations, societies and individuals specializing in the field of population and habitat recovery; however, rarely do these efforts focus specifically on plants of direct socio-economic importance. In general, recovery techniques are reserved for species that are of conservation concern due to their rare or threatened status, combined with their perceived recreational or sometimes intrinsic value, or for unique habitats (although habitat restoration is often carried out for economic reasons, such as protection of watershed or control of soil erosion, in which case the term 'rehabilitation' is often used (see Section 5.1.2)).

Here, we present the argument that recovery techniques can be as applicable for the *in situ* genetic conservation of CWR as they are for those rare and threatened 'pandas' of the plant kingdom. In fact, some of these are CWR and/or dominant species which need to be restored or managed to ensure the conservation of neighbouring endangered plants. Further, since we now know that a high proportion of the flora of Europe and the Mediterranean consists of socio-economically important plants (crops, wild harvested medicinal and aromatic plants and their wild relatives) (Kell *et al.*, 2007a), and by assumption that the same is true for the world flora, the chances of existing recovery actions already including CWR species are also high. There are therefore two ways that population and habitat recovery may be relevant in the context of *in situ* conservation of CWR: (i) there may be a need for the application of recovery techniques as part of the reserve management plan; and (ii) there are undoubtedly existing recovery initiatives involving CWR that can be investigated both as a means of establishing which taxa are already actively under conservation management and of learning and gaining insight from these initiatives.

The need for population and habitat recovery is encapsulated in the text of the Convention on Biological Diversity (CBD), under Article 8 '*In situ* conservation', which states that each contracting party shall, as far as possible and appropriate, 'rehabilitate and restore degraded ecosystems and promote the recovery of threatened species *inter alia*, through the development and implementation of plans or other management strategies'. Dulloo *et al.* (Chapter 2, this volume) present a methodology for locating and designing genetic reserves. In ideal circumstances, reserves are established at sites containing the 'best' CWR populations (i.e., those that best represent the genetic variation present in the species, that provide a suitable habitat for the populations' survival and that are sustainable in the long term). However, in reality, biological and sociopolitical constraints can frequently compromise the ideal genetic reserve. Such constraints may include: a target species that has a very narrow distribution and/or consists of one or only a few populations, limiting the choice of sites for reserve establishment; a species that is represented by one or a few very small populations; the existence of an invasive exotic species at the available, or otherwise ideal, site; the absence of the species' pollinator or dispersal host; the lack or non-recoverable levels of mycorrhizae needed for the target species (e.g. for most terrestrial orchids); issues of land ownership and current resource exploitation; and lack of political or financial backing for reserve establishment at the ideal site. In these circumstances, recovery techniques may be required as a component of the genetic reserve management plan – in fact, they may be the only option.

Recovery techniques can be costly, and their application in plant genetic conservation should be considered with care (Falk and Holsinger, 1991; Maxted *et al.*, 1997). However, recovery can range from the relatively inexpensive and simple process of excluding grazing animals to the resource-hungry process of manual, mechanical or chemical control of invasive species, to the often highly expensive course of collecting, propagating and reintroducing individual plants to enhance an existing population or establish a new population at a site. Therefore, the application of recovery techniques should not be ruled out on the basis of cost before a cost assessment is made; the advantages may easily outweigh the initial outlay, and since monitoring is a required component of reserve management, monitoring of the recovery process can be built into an existing or novel management plan.

Researchers and managers must concentrate their efforts on the detection of true problems for plant conservation and the effective actions needed to solve them, both for species and their habitats. For example, in the case of the Valencian endemic, *Silene diclinis* (Lag.) M. Laínz, it was thought that direct anthropic pressures led to the decline of natural populations. However, more recent studies demonstrated that there were more important problems: seed predation by ants and 'nectar thieves' – insects that perforate the base of the flower to access the nectar but, in doing so, do not pollinate it. It is thought that these insects were displaced from nearby abandoned fields converted to citrus groves. Conservationists must focus their activities on a pre-established good model whose basis is the best knowledge of the species, and in most cases on anthropogenic or other influences on the ecosystem (Mitchell *et al.*, 1990; Ballou *et al.*, 1995; Verhoeven, 2001). Experience shows that managers of long-term projects can be driven in the wrong direction when the 'tools' (wildlife conservation techniques, as defined in manuals such as Schemnitz (1980) or Bookhout (1994)) are progressively converted into goals. This can occur, for example, when contracts are established with very specialized workers or local naturalists and managers are forced to successively re-contract them year after year. It is important to draft or follow comprehensive models, such as the guidelines proposed by the Society for Ecological Restoration International (SER) (see Section 5.2 and Box 5.10), which are useful both for species and habitat recovery programmes. The definition of clear specific goals based on a previous assessment of the target population is essential to focus the recovery project and steer it in the right direction.

5.1.2 What does 'recovery' mean?

Before looking at the techniques and options available for population and habitat recovery, it is important to know what we mean when we use the term 'recovery' in this context. Definitions of the word 'recovery' include: 'possibility or means of recovering or being restored to a former, usual or correct state', 'the action or an act of returning to a former, usual, or correct position' and 'restoration or return to a former, usual, or correct state or condition, as health, prosperity, stability, etc.' (Oxford University Press, 1993). In the context of plant genetic diversity conservation, 'recovery' is broadly used to refer to the act of assisting populations of plant species or habitats in the process of returning from a non-self-sustaining (or unstable) state to a self-sustaining (or stable) one (e.g., see general approaches in Given, 1994).

Population recovery techniques can generally be categorized as reinforcements, translocations, reintroductions or conservation/benign introductions (see Box 5.1 for definitions) – one or more of these approaches may be involved in a single species recovery programme. More recent techniques also include the enhancement of populations of pollinators and seed dispersers, and improvement of associated mycorrhizal populations. All these techniques, primarily used to aid the recovery of endangered species, can be used to conserve populations of CWR (which of course may themselves be endangered). Habitat recovery techniques (commonly termed 'habitat restoration' in the literature) can involve a wide range of activities, including: controlled burning or protection from fire, alteration of the natural water regime, artificial perturbation (ploughing, mowing, clear-cutting, etc.), fencing to prevent damage by stock, wild animals, humans and vehicles, introduction of stock, control of invasive species (rabbit control, weed control, etc.), pest and disease control, supplementary planting or replanting or soil amelioration (see Section 5.3 and Box 5.8). Habitat-focused restoration usually involves a higher number of plant species than a species recovery programme (e.g., see Laguna, 2003; Bokenstrand *et al.*, 2004).

Other terms found in the literature that may be used in the context of population and habitat recovery include: 're-establishment', 'restitution', 'reinstatement', 'enhancement', 'augmentation', 'revegetation', 'rehabilitation', 'reclamation', 'reconstruction', 're-creation', 'introduction', 'rescue' and 'salvage' (see reviews, references and definitions in Jordan *et al.*, 1987; Krebs, 1989; Brown and Lugo, 1994). There is not always a clear distinction between the meanings of the terms used, but as long as the aims of the project are clearly defined, the terminology is not critical. There is, however, an important distinction to be made between 'restoration' and 'rehabilitation': the aim of 'restoration' is primarily focused on conserving biodiversity, while the aim of 'rehabilitation' is usually primarily functional (e.g., soil stabilization or protection of watershed) (see Box 5.1 for the SER distinction). However, sometimes the two aims can be addressed in tandem.

Box 5.1. Definitions of relevance to population and habitat recovery that have been proposed by individuals or organizations specializing in the field.

Reintroduction
– 'An attempt to establish a species in an area which was once part of its historical range, but from which it has been extirpated or become extinct ('Re-establishment' is a synonym, but implies that the re-introduction has been successful).' (IUCN, 1998)
Translocation
– 'Transfer of material from one part to another of the existing range of a species, either to existing or to new sites.' (Akeroyd and Wyse-Jackson, 1995)
– 'Deliberate and mediated movement of wild individuals or populations from one part of their range to another.' (IUCN, 1998)
Reinforcement/supplementation
– 'Addition of individuals to an existing population of conspecifics.' (IUCN, 1998)
Conservation/benign introductions
– 'An attempt to establish a species, for the purpose of conservation, outside its recorded distribution but within an appropriate habitat and ecogeographical area. This is a feasible conservation tool only when there is no remaining area left within a species' historic range.' (IUCN, 1998)
Introduction
– 'An attempt to establish a population in a site where it is not known either to occur now or to have occurred in historical times, but which is within the known distribution range and habitat type of the taxon. If it is suspected that the taxon may have occurred at a site, but this is not confirmed, then a translocation to that site should be treated as an introduction.' (Vallee *et al.*, 2004)
Ecological restoration
– 'The process of assisting the recovery of an ecosystem that has been degraded, damaged or destroyed.' (SER International Science and Policy Working Group, 2004)
Rehabilitation
– 'Rehabilitation shares with restoration a fundamental focus on historical or pre-existing ecosystems as models or references, but the two activities differ in their goals and strategies. Rehabilitation emphasizes the reparation of ecosystem processes, productivity and services, whereas the goals of restoration also include the re-establishment of the pre-existing biotic integrity in terms of species composition and community structure.' (SER International Science and Policy Working Group, 2004)
Habitat restoration
– 'To put back what has been there at some earlier time.' (Atkinson, 1990)
Restored habitat
– 'A system whose structure and function cannot be shown to be outside the bounds generated by the normal dynamic processes of communities and ecosystems.' (Simberloff, 1990)

For some habitats, other specific terms are used, such as 'afforestation' or 'reafforestation', where the main practice is the plantation of the dominant species. The term 'mitigation' may also be used, but is reserved for cases where the deliberate destruction or degradation of a habitat (e.g., to make way for a new road or building development), is followed by attempted reconstruction of the habitat. This can involve literally 'lifting' or translocating parts of habitats or populations and moving them to an alternative site (e.g., see Box 5.2). The nomenclature of

Box 5.2. Translocation of *Narcissus cavanillesii* A. Barra & G. López as a mitigation procedure in the construction of the Alqueva dam in Portugal. (From Rosselló-Graell *et al.*, 2002a,b; Draper *et al.*, 2006a,b.)

Narcissus cavanillesii A. Barra & G. López is a priority species (Habitats Directive –92/43/CEE) occurring in the Iberian Peninsula and North Africa. From a CWR perspective, an interesting feature of this species is its ability to flower in autumn, as most *Narcissus* spp. flower in spring. One of the only two existing localities of this species in Portugal was bound to become extinct due the construction of the Alqueva dam at the Guadiana basin (Alentejo region). Therefore, a translocation action was planned and implemented from Lisbon Botanical Garden to move the population that was going to be flooded to a safe place. The selection of the new location for the population was made through the establishment of a Geographic Information System (GIS) and the development of a habitat suitability model for the species. The criteria used to select the location included the ecological range of the species, proximity to the original location, land use and protection status of the land. As the original population was spatially structured in a series of patches, the translocation was performed in such a way that the original spatial structure of the population was maintained. So far, the translocated population is in good condition and the percentage of flowering individuals has already reached the original values existing before translocation. A monitoring programme is under way to assess the dynamics of the translocated population and, if necessary, implement any needed interventions.

recovery techniques is not a point to dwell on, as long as the aims and required management prescriptions of any type of recovery technique are clearly stated in the protected area/genetic reserve management plan. However, it is useful to refer to some definitions of relevance to population and habitat recovery that have been proposed by individuals or organizations specializing in this field (Box 5.1).

It is very important to clarify the focus for protected area/genetic reserve establishment. Typically, a CWR genetic reserve is designed on a species basis, not on a habitat basis, and with a view to conserve the variability within the species; therefore, the recovery of habitat will be developed in the context of the recovery of the target species. This is a critical issue, since most recovery techniques and activities have historically focused on the habitat or at least on the main structural species of the habitat – those being dominant for that habitat giving it their main structural and functional traits. Additionally, the recovery techniques used have tended to be correlated with the approach chosen by the reserve designer and managers (Young, 2000). These range from the species approach to the habitat, ecosystem (e.g., Cox, 1997) and even the landscape approach (Whisenant, 1999). Each of these approaches is linked to particular methodologies and technical languages and can involve different knowledge fields (species recovery, habitat restoration, ecosystem restoration and landscape restoration); however, rarely have they taken the genetic dimension into consideration. The choice of approach has generated ongoing discussions (e.g., see Maltby *et al.*, 1999), but in most cases the specific solutions or techniques to solve each conservation problem must be designed *in situ* at a local level.

5.1.3 The history and driving forces of population and habitat recovery

Population and habitat recovery is not a new topic. For instance, as early as the late 17th and early 18th centuries, colonial administrators had expressed concern about the fate of the rapidly deteriorating environment of the remote Atlantic island of St Helena and the Indian Ocean island of Mauritius, and recommendations were made to replant areas of the forest. These early attempts were driven by anxiety over increasing scarcity of firewood and timber, severe soil erosion and serious risk of drought (Brouard, 1963; Grove, 1995). In Spain, the use of complex techniques to recover coastal dunes, partially using local species, was successfully developed at the beginning of the 20th century (Garcia *et al.*, 1997; Laguna, 2003). However, it is only in the past three to four decades (from the 1970s onwards) that recovery has been more intensively studied and written about extensively in the literature.

'Population recovery' as a discipline frequently remains 'hidden' within broader topics like 'species conservation'. However, site restoration and management (restoration ecology, restoration biology, etc.) has been progressively recognized as a scientific discipline (e.g., see Young, 2000; Diggelen *et al.*, 2001; Verhoeven, 2001). The articles and works of some of restoration ecology's pioneers like Bradshaw (1983), Bradshaw and Chadwick (1980) and Dodson *et al.* (1997) help us map the evolution of the science of ecological restoration/restoration ecology. Approaches to solve the problems of habitat degradation and loss have generated specific branches of science like 'habitat restoration' (Buckley, 1989; Warren and French, 2000), 'ecosystem restoration' (Taylor, 1995; Samson and Knopf Fritz, 1996; Rana, 1998; Pirot *et al.*, 2000) and 'landscape restoration' (Harker *et al.*, 1990; Makhzoumi and Pungetti, 1999; Whisenant, 1999). During these three to four decades, journals dedicated to restoration ecology, the formation of the SER and the Reintroduction Specialist Group of the IUCN Species Survival Commission, have served to formalize experience and develop a network for sharing information.

Islands, particularly some of the remote oceanic islands, such as Mauritius, Rodrigues, St Helena, Pitcairn and Easter Island (Rapa Nui), have also been at the forefront of studies in habitat and population recovery. Due to the intrinsic vulnerability of island habitats and the rapid degradation of these habitats due to human colonization, islands present unique issues and opportunities for population and habitat recovery and much can be learnt from the last 20–30 years of experimentation. Furthermore, islands are known to harbour unique plant genetic diversity due to the prevalence of endemic taxa and we know that many of these taxa are species of direct socio-economic importance (Kell *et al.*, 2007a); therefore, islands may be important locations for CWR genetic reserves.

Highly modified islands have even been proposed for the partial restoration of a mainland community by providing a habitat free from predators and herbivores (Atkinson, 1990; Simberloff, 1990; Miller *et al.*, 1994). Combined with mammal eradication, species translocations have become a widely used management tool in New Zealand. Similarly, in the Mascarenes, the small offshore islet Ile aux Aigrettes is used as a refuge for threatened lowland native plants of Mauritius

(Dulloo *et al.*, 1997), the coastal ebony forest having been destroyed through clearance for sugarcane plantations. It has also been suggested that Round Island, a small uninhabited island close to Mauritius, be used as a refuge for threatened species from mainland Mauritius, Rodrigues and Réunion (Merton *et al.*, 1989). Could islands present an opportunity as refuges for CWR populations in an uncertain and changing world? With the threat of sea levels rising due to climate change, this approach could be risky for low-lying islands such as Iles aux Aigrettes, but islands with higher elevations may be safe.

Historically, there have been two primary driving forces for population and habitat recovery: economically driven incentives and biodiversity driven incentives. Examples from the wide range of literature on the topic of island habitat restoration serve to highlight these driving forces. Economically driven incentives include watershed conservation and erosion control, such as on the Cape Verde Islands (Lopes and Meyer, 1993) and the provision of commodities such as food, fodder, firewood and timber. Economically driven projects are often classed as 'rehabilitation' (see Box 5.1). Biodiversity driven incentives include conservation of key species and habitats, such as the GEF-funded Mauritius Biodiversity Restoration Project (World Bank, 1995); scientific opportunity, such as the restoration of dry tropical forest on St John, US Virgin Islands (Ray and Brown, 1995); and individual interest, such as the restoration of Nonsuch Island, Bermuda (Wingate, 1985). Other incentives that can broadly be classed as primarily biodiversity driven (rather than economically driven) are legal, cultural, educational, amenity and recreational incentives. Many recovery actions are driven by the perceived value of a species for its sheer beauty and/or intrinsic value (e.g., activities to reintroduce the lady's slipper orchid, *Cypripedium calceolus* L.) (Box 5.3, Fig. 5.1).

In some countries, the development of recovery plans for species and/or habitats has been established on a legal basis. For example, in the context of the US Endangered Species Act, the US Fish and Wildlife Service has extensive experience in recovery projects and the approval of specific, legal rules to protect habitats. In Europe, the Habitats Directive (European Communities, 1995–2007a) has promoted restoration projects focused on biodiversity conservation, including habitat recovery. An example is the Priority Habitats Conservation Project, developed by the Government of Valencia in Spain (Laguna *et al.*, 2003, 2004), where many CWR play a role as important 'structural' species for those ecosystems (Laguna, 2003). Some European countries have developed complex but effective programmes under the umbrella of legal rules and agreements between NGOs, governmental bodies and research centres, whose implementation requires the drafting and development of action plans for species and habitats; for example, the UK Biodiversity Action Plan (JNCC, 2001–2006). Other recent activities have been linked to ecotourism (e.g., see Hammitt and Cole, 1998; Arnberger *et al.*, 2002) and culture (e.g., re-creation of ancient vegetation in archaeological sites, including the reinforcement of endangered species that were abundant in the past, or conservation and rehabilitation of archaeological sites in natural habitats (Forestry Commission, 1995)). Restoration for cultural reasons is a relatively new concept that has recently been proposed by several authors, including Jordan (2003) and Higgs (2003).

Box 5.3. Reintroducing and reinforcing *Cypripedium calceolus* L. at the Ordesa and Monte Perdido National Park (Spain).

The lady's slipper orchid, *Cypripedium calceolus* L. (Fig. 5.1), is one of the most emblematic threatened orchids in Europe. At the Ordesa and Monte Perdido National Park, two localities of this species are presently known. However, only six non-reproductive plants have been scored in one of the localities, whereas the second population, although composed of flowering individuals, is small-sized. As part of a project to promote the long-term viability of the lady's slipper orchid in the park, a habitat suitability model was built based on GIS methodologies to identify suitable sites for the introduction of self-sustainable populations. A total of 16 environmental variables were selected for its significance in explaining the presence of the lady's slipper in the Pyrenees. The territory of the park was also evaluated and scored according to seven factors that condition the successful establishment of lady's slipper populations. By crossing the areas selected through the habitat suitability model with those that obtained the highest score in the evaluation, a total of ten potential sites were identified. In order to select the most suitable source material for the reintroduction, a genetic characterization of all known *C. calceolus* populations in the Central Pyrenees was performed. The genetic study showed that the populations were little differentiated and that source material from both Tena and Pineta valleys would be suitable for the reintroduction at the Ordesa and Monte Perdido National Park. Seeds obtained from these populations are currently being used to obtain plantlets through germination and protocorm development under *in vitro* conditions at the Universidad Politécnica de Madrid. These plantlets will be used in the establishment of new populations. The life cycle of this plant is very long and, in natural conditions, it takes around 11 years for a plant to reach maturity and flower for the first time. Therefore, the monitoring of future reintroductions is likely to extend for many years until the first signs of natural regeneration are observed.

Fig. 5.1. *Cypripedium calceolus* at Sallent de Gállego, Huesca, Spain; one of the sites included in the project to manage and recover the Spanish populations of this species.

Biodiversity driven incentives are usually classed as 'ecological restoration projects' or 'species recovery programmes'. However, there is not always a clear dividing line between economically driven and biodiversity driven incentives. In many cases both the needs of the conservationist and the economic needs of the local community can be addressed in tandem. On St Helena, restoration of the original tree fern thicket was undertaken in the context of watershed security, while in Mauritius, control of the pernicious invasive weed, *Lantana camara*, has benefitted not only native biodiversity, but has also served the needs of local graziers (M. Maunder, Kew, 1997, personal communication). Other initiatives have developed from initial economic incentives to conservation interests. For instance, on the island of Santa Catalina, the primary focus from the 1950s to the 1970s was on the use of the island for livestock grazing (O'Malley, 1991). From 1970 to 1985 there were more concentrated restoration efforts after island managers were awakened to the uniqueness of the flora, and since 1985 more intensive restoration involving planting was driven by the realization that natural vegetation regrowth was too slow to adequately overcome the effects of erosion (O'Malley, 1991). The same joint vision of ecological and economic benefits is also shared by restoration projects focused on coastal or riverine habitats, sometimes centred on the plantation of one or a few species which are simultaneously considered CWR and which can provide direct economic benefits; such as the use of *Populus nigra* L. across Europe (Hughes, 2000; Winfield and Hughes, 2002).

Many projects are focused on the mitigation of the negative effects caused by man and/or on the prevention of damage to property (Saunders *et al.*, 1993; Cairns, 1995; Diggelen *et al.*, 2001). In such cases, the recovery of the habitat or species has not been fixed as a goal, but the result can be very similar. For instance, by the 1950s and 1960s, a plethora of experiences to recover damaged rivers and coastal grounds were developed in industrial countries such as the USA and Japan. These projects focused on the mitigation of damage caused, for example, by the effects of pollution and war and the defence of human habitations against the effect of natural catastrophes such as flooding and coastal erosion. In most cases, these actions can be considered as primarily economically driven. Even though such projects are not designed to directly benefit target species for plant conservation, the activities of re-creation of severely damaged ecosystems – landfills, mined sites, etc. – have generated many new techniques which can be used for CWR or other target species. Interesting reviews of this topic can be found in Andrews (1990), Moffat (1994), Otten *et al.* (1997) and White and Gilbert (2003).

5.1.4 Justifying the use of population and habitat recovery techniques

The decision to use recovery techniques should only be made after all possible questions have been asked about a site or species, and the full range of management options considered. The justification and choice of specific techniques have been analysed both in general reviews (e.g., Given, 1994; Bowles and Whelan, 1994; Frankel *et al.*, 1995; Hambler, 2003) and in the wide range of specific literature for species and habitats. In-depth knowledge of the site is critical and the conservation manager should evaluate the importance of geographical data at large, medium and small scales (e.g., see Morse and Henifin, 1981). The site or habitat where a locally extinct species previously occurred may not be the best site for reintroduction; the reasons for the initial local extinction may still exist at the site (e.g., lack of pollinators, disease or exotic invasive species). Unless the cause of the extinction is known and existing negative influences eliminated, a huge amount of resources could be wasted.

A critical issue to consider in the early stages of genetic reserve planning is that newly created habitat cannot compensate for habitat that has been developing over many hundreds or sometimes thousands of years (Burch, 2001). The complex ecological processes and the functioning of ecosystems cannot be readily recreated by recovery procedures. While some of the component species may be replaced from the start of the programme, many others may not become established until the habitat develops or matures, or until they have recolonized from other sites. Furthermore, the conditions under which the original habitat developed are likely to have been radically altered. In particular, habitat fragmentation may have left sites isolated from other areas of similar habitat, thus significantly reducing the opportunities for recolonization and presenting specific problems for species and habitat recovery (Hobbs and Saunders, 1992; Schwartz, 1997). Part of the skill of the conservationist is to select sites where recolonization is still possible, or to help colonization by introducing propagules or individuals from other nearby habitats. Burch (2001) also points out that there is a danger that in the eyes of developers the offer of an equivalent or greater area of new habitat elsewhere may be seen as a viable choice. In these circumstances, habitat restoration should always be seen as the last resort (Given, 1994). In the case of ecosystems damaged by human activity, preventing habitat destruction or deflecting it to a less sensitive site must be the first priority. There are, however, cases which unavoidably necessitate the complete rebuilding of habitats, such as the effects of volcanic eruptions and landslides. Walker and del Moral (2003) give examples and solutions for the effect of such natural events.

As regards reintroduction, Falk *et al.* (1996) provide practical advice on deciding whether this option is appropriate. By working through a series of questions, the conservation planner can make informed decisions. Such questions include:

- What guidance can be found in existing policies on species reintroductions?
- What criteria can be used to determine whether a species should be reintroduced?
- Is reintroduction occurring in a mitigation context involving the loss or alteration of a natural population or community?
- What legal or regulatory considerations are connected with the reintroduction?

Falk *et al.* (1996) outline some important take-home messages that emerge in existing policies on reintroduction: reintroduction is laden with uncertainty; determining the outcome may take many years, or even decades; reintroduction is largely experimental and presents many opportunities for learning; outcomes of reintroduction, whether positive or negative, should be documented and published; planning and long-term commitment are essential for the success of reintroduction.

In the case of habitats, decisions regarding the best approaches to restoration are often taken in the context of, and as a complementary measure to, a particular species' recovery. There are models available to help decide and plan the recovery techniques to be used (usually referred to as DSSHR – Decision Support Systems for Habitat Restoration), as developed by Pywell *et al.* (1996). However, these models have been established considering the habitat, not the species living in this habitat, as the main target, so their usefulness must be prudently analysed. It is also important to stress that techniques for population and habitat recovery can give contrary results at different sites, depending on habitat type, local history and external pressures on the site, because different processes may be operating in each case (e.g., see Parker, 1995; Lill, 2000). In Spain, the regional administration of Andalucia has protected more than 300 relevant micro-sites against grazing (both by wild and naturalized individuals of domestic goats) in the Nature Parks of Sierra de Cazorla and Sierra Magina, province of Jaen, using successful fencing techniques and saving many endemic species from extinction, including *Atropa baetica* Willk., *Santolina elegans* DC., *Viola cazorlensis* Gand., *Erodium cazorlanum* Heywood and *Andryala agardhii* DC. (Benavente and Luque, 1998) (Plate 10). However, using the same technique in Valencia, eastern Spain, in the framework of experimental studies of the endemic and protected sea lavender, *Limonium dufourii* (Girard) Kuntze, the fencing used to exclude cattle and goats also excluded rabbits, but because rabbits control the growth of plant competitors to *L. dufourii*, such as *Halimione portulacoides* (L.) Aellen, these species finally displaced the protected species (E. Laguna, Valencia, Spain, personal communication, 1997, 2006). In this case, the technical characteristics of the fence make a critical difference.

5.1.5 Establishing recovery goals

Establishing the goals of the recovery programme is a critical part of the management planning process (see also Maxted *et al.*, Chapter 3, this volume). Recovery goals are determined by a wide range of information on the species and/or habitat (Mitchell *et al.*, 1990). The guidelines and comments given in this chapter are built on the basis of reasonable knowledge of the ecogeographic characteristics of the species or habitat and of the reasons causing population decline or habitat degradation. However, frequently, the information needed to make informed decisions is not available in its entirety; therefore, in some cases, the first recovery action is to collate and/or generate the information needed to make further decisions. In such cases, complementary *ex situ* actions are critical (e.g., the regular collection and storage of seeds or spores from the full range of populations of the species).

The ultimate scientific goal of the recovery programme is dictated by a multitude of factors, including habitat type, level of degradation of the habitat and/or populations at the site, species' biology and distribution, minimum viable population size (MVP) and effective population size. However, there are a number of specific problems associated with population and habitat recovery. In extreme cases, habitat destruction may have been complete, with little or no original habitat remaining – this has been the case on many oceanic islands, such as Easter Island and lowland St Helena. In other cases, such as the highly degraded Indian Ocean island of Rodrigues, the earliest known scientific account (Balfour, 1877) was written

200 years after colonization, when already only remnant fragments of semi-natural vegetation existed (Strahm, 1989). In these circumstances, there is frequently a lack of historical data on the original composition and ecology of the habitat; therefore, the goals of habitat recovery must be based on an interpretation of ambiguous historical records and surviving, but often modified, habitat fragments (M. Maunder, Kew, 1997, personal communication). Archive material can, however, reveal important information for recovery planning. Correspondence, maps and plans, sketches and paintings, travelogues and diaries, herbarium specimens, accession books for botanic gardens, and collecting books allow the manager and scientist to develop a historical perspective on degraded habitats and lost species. For example, historical descriptions have been used to assist in the development of species and habitat programmes on Easter Island, the Mascarenes and, most notably, St Helena (M. Maunder, Kew, 1997, personal communication). While oceanic islands often present extreme cases of habitat degradation, coupled with little detailed historic knowledge of the biodiversity of the area, extreme levels of degradation and lack of detailed knowledge of the habitat also occur on mainland areas. Much can be learned from population and habitat recovery projects that have taken place on islands, where a wide range of techniques have been applied and experimented with, particularly during the last 20–30 years.

In other cases, habitat destruction may be less extreme, but the original ecological communities may have been severely disrupted by forces such as invasive species and human pressure. In these situations, detailed studies such as those made by Cronk (1989), Allen and Wilson (1991), El-Demerdash (1996), Kell (1996) and Williams (1997) can help establish recovery goals. By using multivariate analysis programmes such as Two-Way Indicator Species Analysis (TWINSPAN) (Hill, 1994), Detrended Correspondence Analysis (DECORANA) (Hill, 1994) and Numerical Taxonomy System (NTSYSpc) (Rohlf, 1998), data collated from vegetation surveys, historical records and relict occurrences can be used to aid in the definition of species composition for vegetation communities. Palynological studies are also a potential aid to ecological restoration. A study on Mana Island, New Zealand, revealed that pollen analysis of soil and sediment could yield information about pre-settlement vegetation (Chester and Raine, 1990). On Easter Island, fossil pollen records indicate the existence of forests prior to colonization by man–the records showed evidence of a number of woody species, including a now extinct palm (Flenley and King, 1984).

Invasive species are a major issue for both island and mainland populations and their habitats (Pyusek *et al.*, 1995; Williamson, 1996; Luken and Thieret, 1997; Sandlund *et al.*, 1999; Mack *et al.*, 2000; Dulloo *et al.*, 2002; Mooney and Hobbs, 2002; Veitch and Clout, 2002; Davis, 2005; Mooney *et al.*, 2005). Mainland studies of invasive species have primarily focused on fragile and sensitive ecosystems, such as dunes and wetland habitats (e.g., Waal *et al.*, 1994; Howard, 2000; Leppakoski *et al.*, 2002). A long list of invasive species are having a significant impact throughout Europe (de Klemm, 1996; Lambinon, 1997) and their control has been included as a focal target in the European Strategy for Plant Conservation (Smart *et al.*, 2002) and by the Council of Europe (Genovesi and Shine, 2003). In addition, global strategies (McNeely *et al.*, 2001) and guidelines (ISSG, 2000; Shine *et al.*, 2000; Wittenberg and Cock, 2001) have been proposed.

Although the main action to protect an endangered species can be to control or eradicate an invasive species, the conservationist should not forget the main objectives of the recovery programme and the frequent need for parallel actions. After eradication or control, the composition and functioning of the habitat can change dramatically and the same local species can show uncommon population behaviour for many years. A practical example of conflicts arising between CWR species and invasive species can be found on the island of Menorca, Spain. In July 2001, the European Commission approved the funding of a LIFE-Nature project proposal focused on the conservation of endangered plant species in Menorca (LIFE2000NAT/E/7355). The project sought to achieve its objectives by means of three methods: legal tools (recovery plans), control of threats and awareness-raising. The presence of the alien invasive plants, *Carpobrotus edulis* (L.) N.E.Br. and *C. acinaciformis* (L.) L.Bolus, posed a major threat to the conservation of the endangered flora; therefore, a significant number of measures within the project focused on their control and eradication. Among the endangered plant species threatened by the colonization of these invasive plants was *Vicia bifoliolata* Rodr. This CWR legume is a narrow endemic that has a single location in east Menorca. It is a small climbing plant that grows on native shrubs that were being displaced by *Carpobrotus* spp. The successful control of *Carpobrotus* spp. in Menorca has prevented the extinction of this species.

A further problem associated with recovery planning is that major ecological processes are likely to have been disrupted or destroyed over time; for instance, colonial settlement on the island of St Helena was accompanied by periods of severe soil erosion and the loss of massive sea bird colonies has changed the nutrient status of the island's terrestrial ecosystems (M. Maunder, Kew, 1997, personal communication). Accordingly, habitat recovery must often proceed in highly modified edaphic environments. Strict restoration goals can also be frustrated through the unavailability of extinct taxa. For instance, on the island of Rodrigues, two species of giant land tortoise (*Geochelone* spp.) were once present in flocks of 2000–3000 (Gade, 1985). These and a number of other now extinct members of the Rodrigues biota presumably played a major role in the island's ecology and cannot be reinstated. In response to this problem some authors have suggested the use of species substitutions in habitat restoration (see Werner, 1987; Atkinson, 1988, 1990; Dulloo *et al.*, 1997). Trials using the closest relative of two extinct species of Mauritian giant tortoises, the Aldabran tortoise *Geochelone gigantea*, as a substitute for the original large herbivores on the offshore islet, Iles aux Aigrettes, have proven positive (IUCN, 1995–2006; Eskildsen *et al.*, 2004) (Plate 6). This specific problem seems to play a more relevant role in island ecosystems, where the function of a local species cannot be easily substituted by another if it becomes extinct. On mainland areas, the problem of substitution with large herbivores may be easier (e.g., the case of Saiga antelope in the Russian steppes (Struchov and Kuleshova, 2005), American buffalo (Callenbach, 1995; Matthews, 2002) or European bison (Zdzislaw and Belousova, 2004)). In general, the disappearance of smaller animal species on mainland areas is less significant because in most cases their function is quickly replaced by other local competitors.

Whatever the ultimate aim of the CWR recovery programme, the management goals or objectives should be clearly stated in the management plan (see Maxted *et al.*, Chapter 3, this volume). The two case studies presented in Boxes 5.4 and 5.5 provide examples of recovery projects with clearly stated management objectives. As for all management prescriptions, those that are implemented as part of the recovery process should be carefully monitored over

Box 5.4. Ile aux Aigrettes, Mauritius: restoration goals. (From Dulloo *et al.*, 1997.)

The vision

The ultimate aim for Ile aux Aigrettes is to have a self-sustaining indigenous ecosystem with all the components of its flora and fauna living together in harmony in a fully restored habitat. This will serve as a model for ecological restoration work elsewhere in the world. It will also serve the needs for a growing ecotourism industry and act as a living laboratory for the instruction of Natural Sciences to students and the public at large.

Management objectives

To attain our goal, it will require years and years of hard work and resources. To make our tasks simpler, we have set ourselves targets for the short, medium and long term.

In the short term, i.e. in the next 5 years, we plan to resolve many of our current weed problems and barriers to the restoration work, as well as to obtain baseline data and improve our knowledge on the biological components of the island. We need to find out the most cost-effective ways of reversing the degradation of the habitats on the island and to start the revegetation in the most degraded parts while enhancing the good core areas of native forest. This will be achieved through developing appropriate infrastructure for restoration work and strengthening the capacity of the Mauritian Wildlife Foundation (MWF) in project planning and administration, horticulture, conservation biology and project monitoring, involving as far as possible Mauritian staff and students.

In the medium term, we aim at restoring a well-structured native plant community, involving the expansion of the revegetation work in more degraded areas and increasing native species diversity, both of plants and animals. We hope that in the medium term, we would have developed restoration techniques that would be applicable elsewhere. Ile aux Aigrettes would be seen in a wider perspective integrating terrestrial conservation work with the marine environment. An integrated approach to conservation of the region as a unit will enhance the quality of the area and of its local inhabitants. To make our efforts sustainable, resources will be generated through local and international funding and ecotourism.

The longer-term objective, i.e. beyond 20 years, is to re-establish and manage a native ecosystem. The coastal forest will be completed as far as possible with missing components (both plants and animals), which were once thought to exist on Ile aux Aigrettes, over 400 years ago, or suitable ecological analogues. Management of Ile aux Aigrettes will be the 'flagship' of an overall coastal zone management strategy for Mauritius. It will serve as an opportunity for long-term research and training, and as a model for other habitat restoration and species recovery efforts locally, regionally as well as internationally.

Box 5.5. Columbretes Islands Nature Reserve, Valencian Community, Spain. (From Laguna and Jiménez, 1995.)

The vision

The main goal is the regeneration of the damaged landscape of the 8 ha island, Columbrete Gran (or Illa Grossa) of the Nature Reserve of Columbretes Islands – 14 ha of land and 4000 ha of marine habitat – following a step-by-step regeneration model, with the conservation of the most important endemic species, such as *Lobularia maritima* subsp. *columbretensis* R. Fern. and *Medicago citrina* (Font Quer) Greuter.

Management objectives

A mixed recovery programme for habitats and rare or endangered species was started in 1994, including the control of invasive species (mainly of *Opuntia ficus-indica* (L.) Mill.), the establishment of a nursery and the production of the most important taxa for each recovery step: grassland, salt scrubland, primitive maquis and phrygana vegetation. In 1987, an intensive action to eradicate rabbits (introduced c.1860) was set up successfully. More than 20 species were used for the habitat recovery programme, including the locally threatened endemics, planting approximately 10,000 individuals. Additionally, annuals and short-lived perennials were reinforced by planting seeds, e.g. for *Lavatera mauritanica* Durieu and *Beta patellaris* Moq.

Since 1997, most actions have focused on the recovery of the local Luzerne tree (*Medicago citrina* (Font Quer) Greuter), a tall bush suffering the attacks of the cottony cushion scale (*Icerya purchasi* Maskell). This insect is a plague of the sweet orange trees, growing on citrus crops on the close-by mainland of the Valencian Community. Its control on the mainland was enhanced by the introduction of its depredator, the Australian ladybird beetle (*Rhodolia cardinalis* Mulsant) during the 1920s. However, the appearance of a new plague in the Valencian mainland by 1996, the orange-tree leaf caterpillar *Phyllocnistis citrella* Stainton, was massively combated by local farmers using pesticides, which suppressed the former populations of *Rhodolia*. This inappropriate use of pesticides yielded an outstanding overpopulation of *Icerya*, which reached the Columbretes archipelago – 60 km from the Valencian coast – travelling on the bodies and feathers of migrant birds. After some time there, 400 specimens of *M. citrina* – more than 40% of the world population – were damaged during the spring of 1997 and most of them died in a few days (see also datasheet on *M. citrina* in Montmollin and Strahm, 2005).

Nowadays, the management programme includes a regular introduction of small stocks of *Rh. cardinalis* in Columbretes. The populations of *M. citrina* and other significant species of Columbretes are being researched through intensive studies (e.g., Juan *et al.*, 2004).

The management programmes have been significantly influenced by the results of scientific monitoring. A major issue was the study of pollinators for *M. citrina* and other relevant species – also for other Valencian archipelagos like the Tabarca Islands – which demonstrated that the main useful insects, mainly Syrphidae and related dipterans (see Perez-Bañon and Marcos, 1998), used to have larval phases living on the stems and fruits of *Opuntia maxima*, an exotic plant living in Columbretes at least since the 17th century.

time and the management actions adapted as necessary. The management of a genetic reserve is complex and multifaceted, and when part of the goal of the management plan is population and/or habitat recovery, management and monitoring become even more complex, with many more layers to address.

5.2 Population Recovery Techniques

5.2.1 Overview of population recovery techniques

Although the CWR conservation manager can apply management techniques that have been used in the past for other recovery programmes – perhaps for a species or habitat with similar characteristics – as already highlighted, no one programme will be the same as another due to the complexity of habitats, ecological communities and the biology of the species themselves. Therefore, careful management planning to ensure that the goals of the recovery programme can be met using the appropriate techniques is critical (see Maxted *et al.*, Chapter 3, this volume). Furthermore, inherent in any conservation management activity, including recovery programmes, is the need for adaptive management. Adaptive management incorporates research into conservation action. Specifically, it is the integration of design, management and monitoring to systematically test assumptions in order to adapt and learn (Salafsky *et al.*, 2001).

Population recovery techniques can be broadly classified into two types: intervention techniques and control techniques. The main intervention techniques fall into three categories: reintroduction, reinforcement and translocation. Reintroduction is the establishment of a population in a historical location where the taxon is known to have occurred in the past. Reinforcement is the artificial establishment of new individuals in a natural population to increase its size, with the aim of establishing a self-sustaining or stable population. Translocation involves the movement of a natural population from its original location to a new site. (See Box 5.1 for published definitions.)

Intervention can also involve related activities such as the improvement of reproductive or germination success of the existing population by means of hand pollination or the addition of facilities for a local disperser (e.g., nesting boxes). Some more recent techniques (still under experimentation) include the improvement of the habitat for mycorrhizae (see Barea *et al.*, 1996). The role of microorganisms in plant conservation is an emergent topic that will generate future new techniques (e.g., see Sivasithamparam *et al.*, 2002).

Control techniques include various management practices at the site level, without actively interfering with the population itself. Such techniques include removal or exclusion of introduced mammals, weed and grazing control, artificial burning and cutting, soil amelioration and pest and disease control. These site-level control measures are common to both population and habitat recovery and are discussed in Section 5.3.

5.2.2 Policies and guidelines for population recovery

A number of organizations have produced policies and guidelines to promote correct practice in the reintroduction and translocation of organisms (Box 5.6). These should be consulted when formulating the recovery plan as part of the genetic reserve management plan. General guidelines can also be found in plant conservation texts (e.g., Given, 1994; Wyse-Jackson and Akeroyd, 1994). The Reintroduction Specialist

Box 5.6. Guidelines for population recovery.

– Akeroyd, J. and Wyse-Jackson, P. (eds) (1995) *A Handbook for Botanic Gardens on the Reintroduction of Plants to the Wild*. Botanic Gardens Conservation International, Kew, UK. Available at: http://www.bgci.org/worldwide/occasional/
– Environment Australia (2002) *Revised Recovery Plan Guidelines for Nationally Listed Threatened Species and Ecological Communities under the Commonwealth Environment Protection and Biodiversity Conservation Act 1999*. Available at: http://www.deh.gov.au/biodiversity/threatened/publications/recovery/guidelines/index.html
– Falk, D.A., Millar, C.I. and Olwell, M. (1996) Guidelines for developing a rare plant reintroduction plan. In Falk, D.A., Millar, C.I. and Olwell, M. (eds) *Restoring Diversity: Strategies for Reintroduction of Endangered Plants*. Island Press, Washington, DC, pp. 453–490.
– FWS (2004) *Endangered Species Related Laws, Regulations, Policies and Notices*. US Fish and Wildlife Srevice. Available at: http://www.fws.gov/endangered/policies/index.html (Includes a list of policy documents for species and habitat recovery)
– FWS (2006) *US Fish and Wildlife Service Threatened and Endangered Species System (TESS)*. US Fish and Wildlife Service. Available at: http://www.fws.gov/endangered/policy/index.html (Provides access to an extensive list of species and population recovery plans)
– IUCN (1998) *Guidelines for Re-introductions*. Prepared by the IUCN/SSC Re-introduction Specialist Group, IUCN, Gland, Switzerland/Cambridge. Available at: http://www.iucn.org/themes/ssc/publications/policy/reinte.htm
– Vallee, L., Hogbin, T., Monks, L., Makinson, B., Matthes, M. and Rossetto, M. (2004) *Guidelines for the Translocation of Threatened Plants in Australia*, 2nd edn. Australian Network for Plant Conservation, Canberra. Available at: http://www.anbg.gov.au/anpc/books.html

Group (RSG) of the IUCN Species Survival Commission is also a useful reference point for information on population recovery (see RSG, no date).

Practitioners should also consult any relevant national or regional policy or legislation that may be applicable. For example, this could include legislation concerning the movement of germplasm, endangered species acts or existing species action plans. The web site of the appropriate government department responsible for agriculture and the environment should be consulted.

5.2.3 Devising population recovery plans for individual species

The policies and guidelines referred to above provide general guidance that must be adapted to the specific needs of individual species and circumstances. The management planning phase of genetic reserve establishment is a complex process involving detailed assessment of all aspects of the site and the populations of conservation interest (Maxted *et al*., Chapter 3, this volume). If a population requires intervention with recovery techniques, a further raft of complex management decisions must be made (e.g., see Marrero-Gómez *et al.*, 2003). The current state of the art in the recovery of plant populations is hindered by a lack of clear, step-wise case studies providing protocols that can be adapted for individual species. The RSG has made some progress in the provision of taxon and species-specific guidelines, but so far these are only available for animals (primates and elephants). The plant conservationist must therefore draw on a wide array of information sources in order to fully understand the biology and genetic diversity of the species and to make decisions about management interventions; such as controlled pollination and grazing, weed control, seed collection, propagation, population enhancement, reintroduction or translocation.

The primary information source is likely to be peer-reviewed literature, and searches are greatly facilitated by the recent advances in access to online library catalogues and journals. Web sites and grey literature are also likely to play a major role in information provision – much information exists in the form of management plans and reports of existing programmes. However, gaining access to such literature may be less straightforward and more time-consuming. Large institutions such as the US Fish and Wildlife Service offer a comprehensive web site providing wide-ranging information on species recovery (see FWS, 2007a), including a comprehensive list of species recovery plans, but this level of recovery activity and access to information is uncommon. Taxon and disciplinary experts can also be consulted. Taxon experts can be sought by contacting a local herbarium or botanic garden, or through literature and web searches, where e-mail addresses or other contact details are often given. Disciplinary experts include members of the SER (see SER, no date), Society for Conservation Biology (see SCB, 2005) and scientists working in research institutions, such as universities, botanic gardens and gene banks.

In some cases, information may not be forthcoming for the taxon of conservation interest. In these cases, the conservation planner must assume the role of detective in making best use of historical and current information on the site and the species. For instance, this may include adapting protocols used for another species that shares similar biological and genetic characteristics. In this case, a degree of trial and error is likely to be involved, since each site and population varies according to different local circumstances.

In addition to species and habitat-related data, information on the local history of the site and, where applicable, former conservation activities must be collated, along with information on local use and value of the species and/or site. This may involve sociological studies or interviews. Additionally, for some species, a parallel project of environmental education focused on the target species can be designed to support the recovery programme; therefore, the opinion and experience of local teachers and other education specialists should be requested. A recovery programme task force can be established involving site managers, wardens, teachers and conservationists, who all play a significant role.

For some species, site management depends on the activities of farmers and other landowners – additional key players in a plant genetic diversity recovery programme. Some countries, such as the USA, Canada and the UK, involve landowners in conservation activities through stewardship schemes (Johnson *et al.*, 1999) (e.g., some of the projects of the Forest Stewardship Council (see FSC, 2003)). A similar scheme directly involving landowners in plant conservation has been developed by the regional administration of Valencia, Spain, under the Plant Micro-reserves Programme (Laguna, 2001a,b, 2004, 2005; Laguna *et al.*, 2002). In such cases, it is critical that the landowners participate in the design – their opinions can be important in the selection of target species, sites and techniques. For example, under the Plant Micro-reserves Programme, the conservation activities of some landowners have been supported by the regional administration of Valencia in order to promote the involvement of other landowners to create networks of private micro-reserves (Laguna, 2001b).

All of the above assumes that the target species has or have already been designated for recovery action. It is not within the scope of this publication to go into detail about the establishment of priorities for conservation action, but this topic is briefly addressed in Section 5.4.

5.2.4 Genetic considerations in population recovery

The plant genetic conservationist is interested in conserving the maximum genetic diversity within and between populations (Falk and Holsinger, 1991; Maxted *et al.*, 1997). The genetic aspects of undertaking a recovery programme cannot and should not be ignored (Ballou *et al.*, 1995). To begin with, an understanding of the reproductive biology of the species is required. If possible, a genetic diversity study of existing populations, including material stored *ex situ*, should be undertaken. However, limited time and resources dictate that this is not always possible; thus, the conservation manager must make decisions based on the available information, as described above. In addition to literature and other information sources, field assessments must be made to gain as much knowledge of the site and population(s) as possible. Environmental factors commonly associated with regional genetic differentiation include aspect, elevation and climate (Knapp and Dyer, 1998). Other factors that affect the degree of genetic differentiation between populations are the species' mating and seed dispersal systems and how populations are distributed across the landscape. Species that are likely to exhibit greater differentiation between populations are self-pollinators, widespread species found in multiple environments and species whose populations are spatially isolated from one another (Knapp and Dyer, 1998).

When embarking on a population recovery programme, key genetic considerations are provenance, the use of genetically variable reintroduction stock and loss of genetic diversity in *ex situ* raised reintroduction stock.

Provenance

The provenance, or source of material used in population recovery, can be controversial. Questions may arise about the availability of reintroduction material, and where that material comes from. In many cases, there may be little choice but to undertake recovery using plant material from non-local sources, when local material is not available or not available in sufficient quantity. However, putting aside the controversy of using non-local material, provenance may in fact be critical to the success of the project, as material from one site may not necessarily adapt to growth in another. This very much depends on the individual species, its distribution and habitat associations. Further, natural populations contain gene complexes that have adapted over time to local environmental conditions. The introduction of, and hybridization with, non-local source material into a population can cause the co-adapted gene complexes to break up, leading to lower fitness – this process is known as outbreeding depression (Lynch, 1991) and can even occur between populations that appear to be adapted to identical extrinsic environments (Lynch, 1996).

However, sourcing the right material for reintroduction can present practical dilemmas on many fronts:

- How local is local? The genetic structure of populations and metapopulations must be thoroughly understood in order to establish the spatial extent of adaptation. Often, the genetic studies needed have not been carried out, are too expensive to carry out or would take too much time.
- Genetic identity versus genetic adaptation – Jones (2003) points out that maximizing genetic identity between the target plant population and the reintroduced plant material does not necessarily maximize genetic adaptation to an altered site; therefore, the use of non-local source material may be necessary.
- Population extinctions – In some circumstances, a population may have gone extinct from an area and the only source material that is available for reintroduction comes from a different population.
- No choice? In the case of severely depleted populations, to introduce sufficient genetic variation, founder individuals for reintroduction stock may have to be sampled from several different sites. Combined with this, historical data are often lacking and conservationists may be starting with little or no knowledge of the species' original locations. In such extreme cases, decisions to carry out reintroduction are based on a great deal of uncertainty.
- Genetic contamination? On severely degraded sites, it could be argued that the reintroduction of non-locally sourced material can do little more harm than has already been done to the environment. However, the suggestion that an area may be genetically contaminated in this way also raises ethical questions.
- Climate change – It has been suggested that the use of local provenance material in recovery and restoration could lead to a 'genetic dead end' that leaves populations incapable of adapting to new bioclimatic envelopes (Harris *et al.*, 2006).

In some cases, knowledge of intra-population genetic diversity can inform the conservationist that intervention is not appropriate, even for very small populations. For instance, ongoing studies with the endemic species *Silene diclinis* in Valencia, Spain, (see Montesinos *et al.*, 2006) have shown micro-familial distributions of the individuals within each population because seed is dispersed over a very short distance by ants. Similarly, genetic studies conducted on the Spanish endemic species *Antirrhinum microphyllum* Rothm. have shown the existence of micro-scale genetic neighbourhoods (Torres *et al.*, 2003). In this case, the creation of new 'safety' populations is preferable, instead of reinforcement of the current ones. For the creation of new populations, seeds can be sourced from the nearest existing populations, but the choice should be made carefully. Populations should be chosen that have fewer barriers to natural dispersers; for example, a river or stream presents a barrier to ants and, therefore, a population relying on ants to disperse seeds that exists in open grassland away from a river or stream is more likely to be successful in the long term. This issue is also very important in the case of bird-dispersed species – the ideal sites to create new populations should preferably be placed on the birds' migratory routes.

For comprehensive reviews of the importance of genetic considerations in population and habitat recovery, see Knapp and Dyer (1998) and McKay *et al.* (2005). McKay *et al.* (2005) also provide a set of recommendations for genetic restoration and Jones (2003) and Jones and Monaco (2007) present a practical approach to choosing plant material for restoration using the 'Restoration Gene Pool Concept'.

Genetic variation in reintroduction stock

As noted by Knapp and Dyer (1998), it is widely agreed by recovery planners that the presence of genetic variation is a critical factor in the success of species reintroductions. Rieseberg and Swensen (1996) note that a transplanted population should include the full range of ecologically adapted variations from a single site in order to increase the chance of preserving the population's full ecological range. The importance of paying attention to genetic variation in reintroduction stock is illustrated by a study of the Hawaiian silversword alliance (*Argyroxiphium* DC. spp.). Rieseberg and Swensen (1996) found that in self-incompatible plants such as the Mauna Kea silversword (*Argyroxiphium sandwicense* DC. subsp. *sandwicense*) it is critical to include progeny from as many different individuals in the source population as possible in order to maximize compatibility among matings and thus ensure high levels of seed set. They found that the low levels of seed set and different growth forms of reintroduced plants were the result of using descendants from only one or two maternal plants. Robichaux *et al.* (1997) conclude that the reintroduction of *Argyroxiphium sandwicense* subsp. *sandwicense* over a period of 24 years has resulted in a population bottleneck due to the use of offspring from only two maternal founders. The authors revealed a 73% reduction in the level of detectable polymorphism between the natural and reintroduced populations. However, Rieseberg and Swensen (1996) also point out that in the case of small, historically depauperate populations, maximizing genetic diversity in a population recovery programme may not always be appropriate because it may lead to loss of critical adaptive features and possibly to outbreeding depression. In-depth knowledge of the biology (including genetics) of the target species is therefore critical.

In order for reintroduced populations to persist, they also need to be large enough to avoid the loss of alleles due to random genetic drift, and increased inbreeding associated with population genetic bottlenecks. Guerrant and Pavlik (1998) provide a review of the practical issues associated with re-establishment of self-sustaining populations of rare plants from *ex situ* collections. One conclusion drawn by the authors is that, to avoid serious genetic problems, reintroduced populations need to be 'quite large'; however, deciding on the population size required in a recovery programme depends on many factors, including the effects of demographic and environmental stochasticity, and metapopulation dynamics.

Ex situ raised reintroduction material: loss of genetic diversity

A further critical consideration is the potential for loss of genetic diversity in *ex situ* cultivated stock. Most reintroduction projects rely on the use of *ex situ* facilities. This is particularly true in the case of rare and threatened species management. For species that are presumed extinct in the wild, the only known material may be held *ex situ*, or, even if wild populations remain extant, they are frequently very small. In this case, the recovery planner faces particular challenges, as he/she is likely to be working from a very limited genetic base. In the worst cases, the only material that is known to exist, whether in one or more *ex situ* facilities, may have been collected from a very limited genetic base. The stock will then have been subjected to storage and/or multiplication in an artificial environment. For example, seeds may be held for long periods in seed banks or they may be propagated and grown in a glasshouse. Whether held in storage or propagated, the material is likely to lose and/or change genetic variation through the selection pressures caused by the storage conditions. The degree of loss of variation depends on many factors, including the inherent biological characteristics of the species itself and the conditions that the material is subjected to in its *ex situ* environment. The ramifications of this general loss of fitness are that on removal from the artificial environment and reintroduction to the wild, the material may simply not survive or adapt to the wild environment, which itself may have changed significantly. The dangers associated with population augmentation using *ex situ* cultivated stock are outlined by Lynch (1996). The author notes that the long-term deleterious consequences of population augmentation (or supplementation) may outweigh the short-term advantage of increased population size. As discussed above, the introduction of non-local source material has the potential to disrupt local adaptations. In particular, some plants show adaptive divergence on spatial scales as little as a few metres (Lynch, 1996). Lynch points out that this problem takes on added significance when the stock has been maintained *ex situ*, as selection over several generations in cultivation is likely to lead to the expression of maladapted genes.

Plants cultivated *ex situ* may also be more susceptible to pests and diseases, and extreme environmental conditions. Further, they may have developed resistance to diseases that are uncommon in the wild. Supplementation with such stock can lead to the unintentional introduction of pests or diseases that will attack and possibly destroy wild populations (Lynch, 1996). An additional factor with regard to *ex situ* material used for reintroduction is hybridization. Material that is cultivated *ex situ* over several generations cannot always be assumed to be pure; the risk of contamination with pollen of a different population of the same species, or a different species of the same genus that is being cultivated at the same time in the same institution, may be high if proper isolation methods are not correctly applied.

5.2.5 Monitoring population recovery

Monitoring is a key element of population recovery, and especially critical in the case of rare or endangered plants (Primack, 1998). Without monitoring, the conservation manager cannot know whether the management interventions have been a success, or whether changes are required in the management plan. Clearly, the cost implications of population recovery have to be fully considered; projects are only sustainable if ongoing funding is secured. Many recovery programmes are implemented with initial funding secured to undertake the initial set-up activities, but fail simply because insufficient resources were allocated to monitoring and evaluation over subsequent years. In designing recovery programmes, planners should define realistic time frames needed to measure the outcomes and incorporate the associated costs into the budget, thus providing a more accurate assessment of the long-term costs. Detailed information on population monitoring can be found in Iriondo *et al.* (Chapter 4, this volume).

5.3 Habitat Recovery Techniques

5.3.1 Overview of habitat recovery techniques

Habitat recovery, or habitat restoration, can involve many different techniques which may be implemented as individual measures, or, more usually, as part of a combined programme. As already mentioned, typically, a CWR genetic reserve is designed on a species and population basis, not on a habitat basis; therefore, the recovery of habitat will usually be developed in the context of the recovery of the target species. A wide range of techniques from the different branches of restoration ecology can be used, but the choice of a technique or a combination of techniques depends on the particular circumstances at individual sites. For example, some methods are appropriate for the repair of severely damaged sites (e.g., overgrazed areas, areas damaged through overdevelopment) and in these cases some planting (revegetation) is likely to be required. For example, a major project has been ongoing for 25 years in the coastal dunes of Valencia, Spain, to recover the coastal vegetation following destruction of the habitat in the 1970s to develop a tourist resort (Box 5.7, Plate 9). The techniques used in these cases would not necessarily be appropriate in less degraded sites where, for example, a soil seed bank may exist, which may mean that no planting is required (see Leck *et al.*, 1989). However, in both cases, intervention may be needed to modify or control invasive plant species or the direct effect of humans and livestock.

Conservation managers should consult the available literature on the habitat type and/or target conservation taxon or taxa as a starting point when planning a habitat recovery programme. However, one of the problems that practitioners will find is the lack of a homogeneous vocabulary for habitats and/or the existence of many local names that can differ between countries and regions. Even common names such as 'grassland' or 'pasture' can have different meanings for different managers using the same language. For habitat names in Europe, the European Nature Information System (EUNIS) classification (Davies and Moss, 1998) can be used as a standard (see EEA, 1993–2007). EUNIS combines and adapts elements of former classifications such as CORINE (1991) and the classification of Palearctic habitats (Devilliers and Devilliers-Terschuren, 1996). The EU's Habitat Interpretation Manual (European Commission, 2007) may be helpful for conservation managers in the EU. However, the conservationist may often have to deal with transitional habitats, ecotones and semi-natural sites, where the use of conventional classifications may be unsuccessful.

Box 5.7. Twenty-five years of plant recovery in the coastal dunes of Valencia, Spain.

Since 1982, the environmental office of the municipality of Valencia (Spain) has developed a project to recover approximately 7 km of coastal vegetation (including the physical re-creation of the dunes), destroyed during the 1970s, when the dune forest of El Saler and its associated ecosystem were destroyed in order to build up a large tourist resort (Sanjaume, 1988). For 25 years, the biologists, geographers and engineers working for this project have selected and cultivated up to 120 plant species, and developed propagation and plantation protocols for different scenarios (windy and sheltered sites, first-line and second-line dunes, etc.). Most dunes have been fully recreated in an attempt to imitate their original position and structure using aerial photographs taken during the 1950s and 1960s, and have been planted up with seeds, bulbs and young plants of the appropriate local species (Sanjaume and Pardo, 1991). Both dominant and rare species were used, including some of the most endangered local taxa, such as *Otanthus maritimus* (L.) Hoffmanns. & Link (which had only three specimens remaining in 1982) and *Juniperus oxycedrus* subsp. *macrocarpa* (Sibth. & Sm.) Neilr. (which had only seven individuals remaining).

The definitive establishment of site restoration protocols during the past few years – a project supported by the European Community LIFE-Nature programme – was refined by analysing long-term results of monitored plots, dating back to the first plantations made in the 1980s and early 1990s. The most suitable species and material for out-planting (seeds, bulbs, rhizomes, etc.) were selected based on these results, and more than 2 million plants – those most effective for the recovery of the dunes and the local biodiversity, including up to 27 taxa from the 120 initially chosen – have been planted (Plate 9). *Otanthus maritimus* is a dominant species that currently covers the windward side of approximately 0.5 km of the first-line dunes, and its estimated population is over 10,000 individuals. Other previously rare species, such as *Calystegia soldanella* (L.) R. Br., *Eryngium maritimum* L., *Polygonum maritimum* L. and *Echinophora spinosa* L., are nowadays dominant species in these dunes. In the case of the marine juniper, *J. oxycedrus* subsp. *macrocarpa*, a LIFE-Nature project is ongoing to recover 21.6 ha of optimum habitat, planting thousands of young junipers; former plantations (approximately 1000 junipers planted before 2004) are being regularly monitored by the plant officers at the site. Results of this project and some downloadable documents can be found at ENEBRO (no date). Additional information can be found in Fernández-Reguera (2001).

There is an abundance of specific techniques, introduced in several manuals and handbooks, such as Bradshaw and Chadwick (1980), Jordan *et al.* (1990), Harris *et al.* (1996), Gilbert and Anderson (1998), Urbanska *et al.* (1999), Pirot *et al.* (2000) and Perrow and Davy (2002). Some authors give very practical advice (e.g., Munshower, 1994; Parker, 1995) or try to work on the basis of a habitat approach (Parker, 1995; Sutherland and Hill, 1995). Habitat restoration techniques are also widely published in peer-reviewed journals, such as *Restoration Ecology* and *Ecological Restoration.* Box 5.8 provides a summary of the main habitat recovery techniques that may be applicable in the context of CWR genetic reserve management. The techniques listed can be classified according to type; for example, preventive versus interventional, and direct (affecting the

Box 5.8. Summary of the main techniques for habitat recovery/restoration of relevance to the conservation of plant genetic diversity in protected areas and genetic reserves.

- Eradication or control of introduced mammals (which may cause overgrazing, depredation of seeds, bark stripping and soil nitrification); techniques include:
  - Trapping and relocation;
  - Exclusion with fencing;
  - Culling;
  - Use of rodenticide (e.g., see Donlan *et al.*, 2003).
- Control of invasive, exotic plants; techniques include:
  - Controlled grazing;
  - Controlled burning;
  - Manual removal;
  - Biological control;
  - Ring-barking;
  - Ploughing;
  - Mowing/cutting.
- Control of pests and diseases – chemical and/or biological.
- Revegetation, which may involve:
  - Soil amelioration;
  - Planting structural or nurse species, which may be both fast-growing, non-invasive introduced species and/or native species;
  - Planting out native species, which may involve some form of protection (e.g., tree guards and/or fencing);
  - Allowing natural regeneration (e.g., by controlling grazing);
  - Soil translocation (i.e., soil patches extracted from a site and transferred to another) – this technique is mainly used in the regeneration of wetlands.
- Control or prevention of soil erosion – there are many different techniques used according to climate, soil type and kind of erosion; examples include:
  - Revegetation;
  - Terraces (e.g., see Cracknell, 1983);
  - Retaining stone walls;
  - Ditches and caldeiras (crescent-shaped plant pits) which can be planted with crops and/or trees (e.g., see Haagsma and Reij, 1993; Lopes and Meyer, 1993);
  - Netting (e.g., see Warren and Aschmann, 1993);
  - Windbreaks, using fencing and/or trees (e.g., see Wingate, 1985);
  - Horizontal paraments;
  - Mulching (i.e., adding layers of organic matter on the soil surface to reduce erosion caused by rainfall and to improve soil structure);
  - Gabion baskets for rivers or coastal grounds.
- Soil amelioration, including:
  - Fertilization of poor soils (e.g., planting selected local species of legumes, using chemical fertilizers);
  - Retention of soil moisture using jute netting (e.g., see Warren and Aschmann, 1993);
  - Increasing potential for mycorrhizae and rhizobium;
  - Improvement in soil structure and nutrients (e.g., using locally available organic matter or biosolid pellets). For example, on the 25 ha coral islet of Ile aux

Continued

Box 5.8. *Continued*

Aigrettes, Mauritius, a by-product of the sugarcane industry ('bagasse') has been utilized to improve soil depth and structure prior to planting.
- Enhancement of populations of useful fauna, such as the establishment of pollinator populations, either directly (e.g., using beehives), or through plantation of additional attractive species, or the establishment of dispersers (mainly birds or ants) through artificial nesting techniques. In the case of birds or mammals, the installation of artificial water points or reservoirs can aid the establishment of new populations.
- Control of human access (e.g., fencing or closing the site, usually with explanation panels), or limited access for educational or cultural reasons, or scientific research.
- Habitat translocation (i.e., lifting and transferring 'pieces' of habitat from one site to another). This is usually used in mitigation circumstances where a habitat is to be destroyed or degraded by development work.

site directly) versus indirect (affecting buffer or surrounding areas, but benefiting the site). However, to categorize the techniques in this way is of little semantic importance – the literature on the subject, particularly manuals and practical guidelines, focus on concrete actions or habitat types, as detailed in Section 5.3.2, rather than general 'types' of actions.

A vast array of habitat restoration case studies are available in the literature and on the Internet. Readers are invited to consult the web sites listed in Box 5.10, but searches via the standard online library databases will also reveal a wide range of publications on the habitat type or species of interest. In addition, some multinational entities have published selections of case studies; for example, in the European Union, results of projects funded by the LIFE-Nature programme can be downloaded from the European Commission's web site (European Communities, 1995–2007b).

When carrying out literature searches, one will undoubtedly find an enormous difference between the accumulated experiences that exist for some kinds of habitats (e.g., dunes, wetlands, rivers, mainly managed and restored during the last 30–40 years for bird conservation and/or against erosion to protect human habitations), and a large list of ecosystems or habitats lacking in former experiences. Additionally, the use of plant species for recovery or restoration has often focused on widely distributed species, such as the forestry species, European beech and Scots pine, perhaps due to the tradition of using only a small group of species whose success has been confirmed across multiple sites. Some isolated experiences, such as those of OEC–DIREN (1998) in Corsica and by the Department of Environment of the Generalitat Valenciana (Box 5.9) in the Valencian Community, Spain, have mixed a combination of techniques to simultaneously obtain information on restoration protocols for a long list of habitats and species.

Box 5.9. Case study: the Priority Habitats Conservation Project in the Valencian Community, Spain. (From Laguna *et al.*, 2003, 2004.)

The goal

This project, funded by the LIFE-Nature programme of the European Union and developed by the Department of Environment of the Generalitat Valenciana (Government of the Autonomous Community of Valencia) during the period 1999–2003, aimed to combine technical knowledge, experience and skills to manage and recover up to 17 different habitat types – all of them considered as 'priority habitats' under the legal umbrella of the Habitats Directive (Directive 92/43/CEE) – in the Valencian Community, Spain. The main aims of the project were to establish protocols for seed collection and germination, *ex situ* cultivation of plants, reintroduction or reinforcement of populations and habitat management for around 200 species (mostly structural–dominant or co-dominant species of various vegetation types, including petrifying springs, salt steppes dominated by *Limonium* spp. and Iberian gypsum steppes). For most of the habitats, no prior work of this kind had been undertaken in Spain or in the western Mediterranean area. A further aim of the project was to increase or create populations of the most endangered plant species in the same habitats.

Management activities

Up to 226 simultaneous, permanent plots for planting and management activities were established, covering 996 ha and distributed across 38 Natura 2000 Sites. The full programme involved plantations of 90,440 young plantlets and 39,092 pretreated seeds, belonging to 213 species (48 of them endemic to Spain). Complementary *ex situ* actions yielded 582 new accessions belonging to 329 species for germplasm banks and primary protocols for germination and growth of 170 taxa. In addition, protocols for *in vitro* propagation focused particularly on germination and growth (without clonal propagation) of embryos from immature seeds of recalcitrant-seeded species.

The field plots encompassed complementary actions like signposting and/or setting up information panels. Public access was limited by roping off 474 ha, generating 152 exclusion areas for visitors and using 15.1 km of rope. As much as 119 ha was clear-cut to benefit herbaceous endemic or threatened species and tree removal was carried out in 109 ha of ancient forestry plantations containing trees such as *Eucalyptus* L' Hér. spp. and other introduced, unwanted species. Extraction and control of invasive species – mainly focused on *Carpobrotus edulis* (L.) N.E.Br., *Opuntia maxima* Mill., *Cylindropuntia tunicate* (Lehm.) F.M. Knuth, *Agave americana* L. and *Robinia pseudoacacia* L. – was undertaken over 85 ha. Several new techniques were developed, such as planting out propagules (leaf cuttings of the endemic *Pinguicula dertosensis* (Cañig.) Mateo & M.B. Crespo on petrifying springs) and successful orchid translocation (*Barlia robertiana* (Loisel.) Greuter, *Ophrys speculum* Link, *O. fusca* Link, *Serapias parviflora* Parl. and the endemic *Ophrys dianica* M.R. Lowe, J. Piera, M.B. Crespo & J.E. Arnold).

The programme also involved protection measures, such as the establishment of 49 new plant micro-reserves covering 485 ha and officially protected by 2001–2003, increasing the number of this kind of protected area in the Valencian Community to 230 sites and 1447 ha since 1998. A long list of complementary activities was developed (e.g., environmental education projects and 13 km of educational pathways), involving local people and task forces, NGOs and landowners. Additionally, actions to involve people comprised a public communication campaign, several exhibitions and setting up 580 m^2 of new rocky gardens holding more than 4300 specimens of endemic and other local species. The results of this project are still under analysis, but examples of activities have been illustrated by Laguna (2003).

The use of non-native species in habitat recovery

As described above, many habitat restoration projects come under the broad heading of 'rehabilitation', which can be thought of as the process of improving the structure and functioning of a habitat. Rehabilitation may be thought of as an end in itself, or as a prerequisite to ecological restoration. Many sites are so severely degraded that the focus has to be on amelioration; for example, the initial requirement may be for land stabilization to avoid further erosion. In ideal circumstances, rehabilitation would not be required in a protected area or CWR genetic reserve because the reserve will have been selected on the basis of the habitat being suitable for conservation of the target species. However, in the case of a species with a very restricted distribution, there may be occasions where habitat degradation has been severe and the initial focus could be on rebuilding the foundations of a sustainable habitat that will support the existing CWR populations or planned reintroduction/introduction/translocation. In such extreme circumstances, non-native species are often selected for replanting. The benefits of using non-native species include:

- Fast-growing species can be chosen, whereas many native species take longer to establish.
- They may be chosen for their resistance to the harsh conditions prevailing at the degraded site (e.g., drought, salinity, shallow soils, strong winds and/or resistance to pests and diseases).
- They may provide a fast solution to a dearth of food, forage, fodder or fuelwood for local inhabitants.
- They may be easier to propagate and grow in large quantities, and thus cheaper to produce.
- They can provide a quick fix for problems of land erosion.
- They can act as a nurse crop for the reintroduction of native species.

The use of exotic species as a nurse crop for the regeneration of indigenous vegetation has been investigated by Parrotta (1995), Kuusipalo *et al.* (1995) and Chapman and Chapman (1996). Each study concludes that exotic tree species can be beneficial in reinstating indigenous forest on degraded sites, although caution should be exercised in the choice of species. On the island of Lakeba, Fiji archipelago, *Pinus caribaea* Morelet was planted to control severe soil erosion – a diverse native cover was established under the remaining 10-year-old stands, runoff was reduced and water percolation was greatly increased (Brookfield, 1981). On Nonsuch Island, Bermuda, a temporary windbreak was established around the periphery of the island by planting two fast-growing exotic species, *Casuarina equisetifolia* L. and *Tamarix gallica* L.; these species were chosen because they do not regenerate in Bermuda soils, and could be eliminated easily once they had served their purpose (Wingate, 1985).

However, while the use of non-native species in rehabilitation and restoration is widespread, there are potential problems. For example:

- They may become invasive (e.g., the use of *Acacia nilotica* subsp. *adstringens* (Schumach.) Roberty in Rodrigues (Kell, 1996)).
- A large area of land planted with one species or variety only may be vulnerable to pests and diseases.
- There may be potential for genetic pollution.
- Non-native species may not support the native fauna and microfauna of the site.

Non-native species should therefore be used with caution and the use of indigenous species should always be considered. In Mauritius, the fast-growing indigenous species *Scaevola taccada* Gaertn. Roxb. and *Hibiscus tiliaceus* L. have been successfully used as nurse plants to provide a niche for the re-establishment of endemic plants of Ile aux Aigrettes. In Ethiopia, restoration of *Olea europaea* subsp. *cuspidata* (Wall. ex G. Don) Cif. forest has been aided by planting *Olea* seedlings under native pioneer shrub cover of *Acacia etbaica* Schweinf. and *Euclea racemosa* Murr., which provide shade, conserve moisture and protect *Olea* seedlings from herbivores (Aerts *et al.*, 2007). Jones and Monaco's (2007) 'Restoration Gene Pool decision-making flowchart' provides a useful, practical guide to the conservation manager when making decisions about the plant materials to be used in the recovery project.

5.3.2 Guidelines for habitat recovery

Habitat recovery (or restoration) techniques have been implemented in many areas of the world, have been the focus of many research projects and have been written about extensively in the literature – the manuals and handbooks referred to in Section 5.3.1 are useful for an introduction to their variety and results. There are also several key journals that focus on restoration experience, most notably *Restoration Ecology*, published by SER. As advised when embarking on a population recovery programme, the CWR conservationist should consult existing information sources if habitat recovery is proposed as a component of genetic reserve management (see Box 5.6).

Due to the wide range of habitats and techniques, mostly adapted to specific cases for each threatened taxon or site, it is not easy to find common guidelines for habitat recovery. Only a few countries (such as the USA) have developed habitat recovery programmes on a legal basis to complement the conservation of endangered species. The web site of the Endangered Species Habitat Conservation Planning Programme, developed by the US Fish and Wildlife Service, provides several useful documents (FWS and NMFS, 1996; FWS, 2005). In addition, the FWS publications web page (FWS, 2007b) provides access to useful documents and examples of habitat recovery projects, linked to species conservation programmes.

Some of the techniques listed in Box 5.8 have generated specialist literature, such as the use of fire (e.g., Pyne, 1996; Dyer *et al.*, 2001; Russell-Smith *et al.*, 2002; Abbott and Burrows, 2003; Cary *et al.*, 2003), the establishment of corridors (Hussey *et al.*, 1991; Watt and Buckley, 1994; Ferris and Carter, 2000; Kuijken, 2003) or the eradication of invasive species (Fernández-Orueta and Aranda, 2001; Myers and Bazely, 2003). In the case of invasive species control, some widely distributed species that have caused problems in many countries have generated specific manuals, such as Japanese knotweed, *Fallopia japonica* (Houtt.) Ronse Decr. (see Child and Wade, 2000; Japanese Knotweed Alliance, no date). Some entities, both public and private, have handbooks, manuals and guidelines available for specific activities. Some of the most relevant ones are listed in Box 5.10.

Often, the easiest way to find information pertinent to a habitat recovery project is to search for projects that have been undertaken in a similar habitat type.

Box 5.10. Useful references for habitat recovery information.

- Society for Ecological Restoration International (SER) (available at: http://www.ser.org/): Edits and distributes a wide range of periodicals and book series and maintains a database of projects. The web site provides access to publication lists and restoration guidelines, notably 'Guidelines for Developing and Managing Ecological Restoration Projects' (available at: http://www.ser.org/content/guidelines_ecological_restoration.asp) and the 'SER International Primer on Ecological Restoration' (available at: http://www.ser.org/content/ecological_restoration_primer.asp).
- The European Union's LIFE Unit: The administrative unit in charge of the LIFE-Nature programme, through which conservation project experience in Europe has accumulated since 1992. Publications containing case studies are available at: http://ec.europa.eu/environment/life/publications/lifepublications/index.htm. In addition, the LIFE-Database provides summaries and contacts for hundreds of conservation projects, most of them involving species recovery and/or habitat restoration (available at: http://ec.europa.eu/environment/life/project/Projects/index.cfm).
- Europarc Federation (available at: http://www.europarc.org/international/europarc.html): The European Federation of National and Natural Parks provides books and reports, mainly focused on protected area management and related issues such as visitors and public interpretation.
- British Trust for Conservation Volunteers (BTCV): BTCV Practical Handbooks Series includes specific manuals for tree planting, fencing, dry stone walling, hedging, etc. Most of them are available as texts or summaries via the web site at: http://handbooks.btcv.org.uk/handbooks/index.
- Natural England (available at: http://www.naturalengland.org.uk/): Useful habitat management handbooks can be downloaded from http://www.english-nature.org.uk/pubs/handbooks/. Additionally, the publications and reports of Natural England can be downloaded from http://naturalengland.twoten.com/NaturalEnglandShop/default.aspx. English Nature's (now part of Natural England) practical guides for coastal habitat restoration can also be accessed at http://www.english-nature.org.uk/livingwiththesea/project_details/good_practice_guide/Home.htm.
- Scottish Natural Heritage (SNH) (available at: http://www.snh.org.uk/): The series 'Natural Heritage Management' includes a range of very useful guides for many different habitats and/or conservation techniques (e.g., fencing), mostly available at: http://www.snh.org.uk/pubs/default.asp. Additionally the SNH programme 'Natural Heritage Future' provides a lot of practical case studies and results (available at: http://www.snh.org.uk/futures/Data/index.htm).
- The Council of Europe: Publishes several book series and reports – a list of publications can be consulted at the Council of Europe's online bookshop at: http://book.coe.int/EN/. For planning and management, mostly focused at landscape level, there is a thematic list available at: http://www.coe.int/t/e/Cultural_Co-operation/Environment/Landscape/Publications/01_Publications_COE.asp.
- Royal Society for Protection of Birds (RSPB) (available at: http://www.rspb.org.uk/): The RSPB has a great deal of experience in habitat re-creation and management. The RSPB Document Library can be consulted at: http://www.rspb.org.uk/ourwork/library/.
- European Centre for Nature Conservation (ECNC) (available at: http://www.ecnc.nl/index.html): A publications list is available at http://www.ecnc.nl/Publications/Index_26.html. The ECNC publication list does not include manuals, but it includes synthetic references for global visions (e.g., on conservation financing and integration of biodiversity in several policies) whose reference lists can be very useful.

As mentioned earlier, searching online library databases can be helpful in sourcing relevant publications; however, we recommend that conservation planners and managers review the local, regional or national literature first, if available, because it is likely to contain information relevant to the particular characteristics and management problems of habitats in the locality (and may also be written in a local language and therefore be more accessible to the reserve management team).

The recovery of a habitat, as is the case for populations, must be monitored. This involves defining a list of appropriate indicators and short-, medium- and long-term goals. Much of the literature cited in this chapter includes specific monitoring case studies or techniques, but of course the conservationist or manager must draft a monitoring programme that is appropriate for his/her specific project. The *SER International Primer on Ecological Restoration* (SER International Science and Policy Working Group, 2004) includes nine attributes that should be considered when evaluating the success of a habitat restoration project. In a review of how success is being measured in restoration projects, Ruiz-Jaen and Aide (2005) conclude that while measuring all attributes listed in the *SER Primer* would be ideal, it is unlikely to be realistic. The authors propose that the measurement of the three attributes – diversity, vegetation structure and ecological processes – which are essential for long-term persistence of an ecosystem, should be promoted as a minimum, and that the criteria to evaluate success should be based on a comparison of more than one reference site.

The genetic reserve manager must not forget that the site will not be maintained by itself, even if abandoned. On the contrary, the activities of the people living near the site can influence the evolution of the recovery programme, and in some cases the success of the project may be compromised by permanent conflict (e.g., see references and case studies in Conover, 2001). Stewardship projects and other models of participation of the private sector and local community are recommended to complement the recovery or management plan. If possible, the project should also include educational programmes and/or environmental education projects. Some useful references on this topic include Knight and Landres (1998), Meffe *et al.* (2002) and Shine (1996). In some cases, the conservation programme could involve the recovery of lost human uses, or their simulation (e.g., reintroduction of livestock, as in the Steppe of Le Crau in Southern France), a special issue, which requires the participation of ethnologists and other specialized managers (see Grayson, 2000). Additionally, the communication of the project's aims and results should not be limited to scientific literature, but should include the publication of popular literature, such as the books of Marren (1999), filled with interesting stories of plant and habitat conservation that are accessible to a wider readership.

Finally, the development of a successful recovery/restoration project depends upon the social and political framework (see a specific analysis in Cortner and Moote, 1999), obligating managers to develop specific skills and techniques in public communication, or to contract specialists.

5.4 Conclusions and Recommendations

In this chapter, we have outlined the relevance of population and habitat recovery in the *in situ* conservation of plant genetic diversity (and, in particular, CWR), reviewed the techniques available for population and habitat recovery, and where information and guidance can be sought on the procedures to be followed when planning a recovery programme as part of a protected area or genetic reserve management plan. As already mentioned, in ideal circumstances, population or habitat recovery should not be necessary for *in situ* conservation of CWR taxa. However, in certain circumstances, the adoption of one or more recovery techniques may be necessary. For example, some small, restricted range populations may require enhancement by introducing individuals to the populations, while at the same time, the target populations' habitat may require some form of specific management to 'restore' it to a predetermined state suitable for the long-term success of the recovery programme. Population and habitat recovery may be needed in tandem because, in many cases, the target population may have suffered a severe reduction in number and reproductive success due to degradation and/or loss of habitat. There may also be cases where the ideal site for the establishment of a genetic reserve is not available due to sociopolitical restrictions. Therefore, in some cases, conservationists may be forced to establish reserves in less desirable sites requiring some form of recovery action.

There is a wealth of information available on population and habitat recovery techniques and some of this information is available in various specific handbooks and guidelines. Some major sources of such information are cited in this chapter – we recommend that these sources are consulted before any form of recovery action is undertaken in a protected area or genetic reserve. However, the diversity of habitats and species' ecogeographic characteristics dictates that no one set of procedures can be followed in all cases. Therefore, when recovery action is being considered or planned as a component of a conservation management plan, literature searches to find case studies involving similar habitats and/or species with similar ecogeographic characteristics should be undertaken. Online library databases are extremely helpful in this regard. There are also a number of specific peer-reviewed journals and web sites that can be consulted and searched.

This volume addresses the conservation of plant genetic diversity in protected areas and genetic reserves, with a particular focus on CWR. As already noted, many population and habitat recovery projects have been established to conserve rare and threatened species per se, or degraded habitats recognized for their general value as characteristic of a particular species composition, or perhaps simply to 'green' degraded areas for recreational and wildlife conservation purposes. It is unlikely that many existing recovery programmes are specifically focused on the conservation of CWR for their recognized value as gene donors for crop improvement; however, there is undoubtedly a wide range of projects that focus on species that are wild relatives of crops, but that have not been identified as such in the conservation management plan.

In terms of addressing the conservation needs of CWR nationally, regionally and even globally, we recommend that a priority list of CWR species in need of recovery planning be developed. This is perhaps an activity that could be undertaken by the recently established CWR Specialist Group (CWRSG) of the IUCN Species Survival Commission (see Dulloo and Maxted, 2007). National lists could initially be drawn up, formed into regional priority lists and eventually into a global priority list of species. This activity will depend on some form of prioritization of CWR species for conservation action, possibly as proposed by Ford-Lloyd *et al.* (2007).

A recommendation made by Kell *et al.* (2007a) was to establish which existing conservation initiatives are already contributing to the conservation of CWR by undertaking a simple cross-check between the taxa listed in the conservation database or inventory and a national or regional CWR inventory. Kell *et al.* (2007a) have already begun this procedure using the Catalogue of Crop Wild Relatives for Europe and the Mediterranean (Kell *et al.*, 2005) for example, to establish which CWR species are included in the Habitats Directive and Important Plant Areas. A similar procedure could be followed to establish which CWR species are already included in population (and habitat) recovery programmes, given access to the required taxon data. Recommendations could be made to members of SER and the IUCN/SSC Reintroduction Specialist Group to help 'tag' the CWR species in existing initiatives. Again, this could be a step taken by members of the CWRSG. The purpose of the SSC network is to work together in this way and this is a prime example of how two Specialist Groups could work together for the benefit of species conservation.

Whenever recovery techniques for plant genetic diversity conservation are incorporated and implemented in a protected area or genetic reserve management plan, a good deal of caution should be exercised as regards the chosen approach. Recovery interventions can be costly and resource-hungry; therefore, a thorough and detailed decision-making programme should be followed when planning the interventions. Furthermore, as outlined in this chapter, the potential genetic effects of recovery actions, particularly with regard to the introduction of germplasm to an *in situ* conservation site, should be carefully considered. However, ultimately, as with many conservation actions, in reality some degree of trial and error is likely to be necessary in order for the project to be a success. Trial and error necessitates careful long-term monitoring. No recovery action should be undertaken without a thorough, well-planned monitoring programme in place. As with any form of conservation intervention, whether largely 'passive' or 'active', long-term monitoring and an adaptive management approach are essential prerequisites for a successful project.

A further consideration when planning population and/or habitat recovery for plant genetic diversity that has been highlighted throughout this volume is climate change. The ranges and habitat niches of CWR species will undoubtedly change as the climate warms up over the coming decades. This presents a potentially mammoth complication in the case of population and habitat recovery. Will the resources spent now on substantial interventions pay off in the long run? If populations of the target CWR or other plant species are 'on the move', will efforts in their current habitats be wasted? Should conservationists be looking at the potential future distribution of the species in the light of climate change models and focusing their efforts there as well as on current target species' sites? Will there be any suitable sites available for the target species at all in 20, 30, 40 or 50 years' time? These and other questions related to climate change remain to be answered.

The likely implications of climate change for ecological restoration have been examined by Harris *et al.* (2006), who have called for more consideration and debate to be directed in this area. The authors highlight the impacts of increasing atmospheric CO_2 concentrations, changes in mean annual temperatures and precipitation, and illustrate how many current site-based conservation initiatives may result in failure. They suggest that to build resilience to climate change into restoration projects, a wider range of species may need to be used and a broader landscape perspective that addresses connectivity to allow species movement will be needed. The authors also question the use of local provenance material in restoration projects, suggesting that in some cases we may be 'consigning restoration projects to a genetic dead end that does not allow for the rapid adaptation to changed circumstances that may be needed if climate change scenarios proceed as predicted' (Harris *et al.*, 2006) – food for thought for the future of plant genetic diversity conservation involving recovery interventions.

Finally, the development of a database of population and habitat recovery references applicable to plant genetic diversity conservation in protected areas, and particularly CWR conservation in genetic reserves, would be a useful addition to the general conservation planning toolkit. This could include papers, books and practical manuals and could be broadly categorized according to habitat type, target or characteristic species, type of management intervention and location. A web-enabled, searchable database could be provided, subject to available resources. This could be linked to existing CWR information resources such as the Crop Wild Relative Information System (CWRIS) (PGR Forum, 2005; Kell *et al.*, 2007b) and any available online sources of information on population and habitat recovery programmes. A population and habitat recovery bibliographic database was developed during the 1990s by the IUCN/SSC RSG, Royal Botanic Gardens, Kew, and the World Conservation Monitoring Centre (WCMC) (Atkinson *et al.*, 1995). This could be used as the basis for the development of a new database, updated with recent population and habitat recovery references and then maintained as an ongoing project. Further to this database, the authors of this chapter also have the seeds of a population and habitat recovery planning bibliography available that will contribute significantly to the proposed database.

References

Abbott, I. and Burrows, N. (eds) (2003) *Fire in Ecosystems of South-west Western Australia: Impacts and Management*. Backhuys, Leiden, The Netherlands.

Aerts, R., Negussie, A., Maes, W., November, E., Hermy, M. and Muys, B. (2007) Restoration of dry Afromontane forest using pioneer shrubs as nurse-plants for *Olea europaea* subsp. *cuspidata*. *Restoration Ecology* 15(1), 129–138.

Akeroyd, J. and Wyse-Jackson, P. (eds) (1995) *A Handbook for Botanic Gardens on the Reintroduction of Plants to the Wild*. Botanic Gardens Conservation International, Kew, UK. Available at: http://www.bgci.org/worldwide/occasional/.

Allen, R.B. and Wilson, J.B. (1991) A method for determining indigenous vegetation from simple environmental factors, and its use for vegetation restoration. *Biological Conservation* 56, 265–280.

Andrews, J. (1990) Principles of restoration of gravel pits for wildlife. *British Wildlife* 2(2), 80–88.

Arnberger, A., Brandenburg, C. and Muhar, A. (eds) (2002) *Monitoring and Management of Visitor Flows in Recreational and Protected Areas*. Institute for Landscape Architecture and Landscape Management, Wien, Austria.

Atkinson, I.A.E. (1988) Presidential address: opportunities for ecological restoration. *New Zealand Journal of Ecology* 11, 1–12.

Atkinson, I.A.E. (1990) Ecological restoration on islands: prerequsites for success. In: Towns, D.R., Daugherty, C.H. and Atkinson, I.A.E. (eds) *Ecological Restoration of New Zealand Islands*. Conservation Sciences Publication No. 2., Department of Conservation, Wellington, New Zealand, pp. 73–82.

Atkinson, P.J., Maunder, M. and Walter, K.S. (1995) *A Reference List for Plant Re-introductions, Recovery Plans and Restoration Programmes*. Version 1, September 1995. IUCN/SSC Re-introduction Specialist Group, Royal Botanic Gardens, Kew, Sussex, UK and the World Conservation Monitoring Centre. Available at: http://www.kew.org/conservation/main.html.

Balfour, I.B. (1877) Aspects of the phanerogamic vegetation of Rodriguez, with description of new plants from the Island. *Botanical Journal of the Linnaean Society* 16, 7–24.

Ballou, J.D., Gilpin, M. and Foose, T.J. (eds) (1995) *Population Management for Survival and Recovery: Analytical Methods and Strategies in Small Population Conservation*. Columbia University Press, New York.

Barea, M., Requena, N. and Jimenez, I. (1996) A revegetation strategy based on the management of arbuscular mycorrhizae, *Rhizobium* and rhizobacterias for the reclamation of desertified Mediterranean shrubland ecosystems. *CIHEAM – Options Méditerranéennes* 20, 75–86. Available at: http://ressources.ciheam.org/om/pdf/c20/96605780.pdf.

Benavente, A. and Luque, P. (1998) La gestion in situ de flora amenazada en el Parque Natural de Cazorla, Segura y Las Villas. Conservacion Vegetal 3, 5–6. Available at: http://www.uam.es/otros/consveg/documentos/numero3.pdf

Bokenstrand, A., Lagerlöf, J. and Torstensson, P.R. (2004) Establishment of vegetation in broadened field boundaries in agricultural landscapes. *Agriculture, Ecosystems and Environment* 101, 21–29.

Bookhout, T.A. (ed.) (1994) *Research and Management Techniques for Wildlife and Habitats*, 5th edn. Wildlife Society, Bethesda, Maryland.

Bowles, M.L. and Whelan, C.J. (eds) (1994) *Restoration of Endangered Species: Conceptual Issues, Planning and Implementation*. Cambridge University Press, Cambridge/New York.

Bradshaw, A.D. (1983) The reconstruction of ecosystems. *Journal of Applied Ecology* 20, 1–17.

Bradshaw, A.D. and Chadwick, M.J. (1980) *The Restoration of Land. The Ecology and Reclamation of Derelict and Degraded Land*. California University Press, Berkeley, California.

Brookfield, H.C. (1981) Man, environment, and development in the outer islands of Fiji. *Ambio* 10(2–3), 59–67.

Brouard, N.R. (1963) *A History of Woods and Forests in Mauritius*. Government Printer, Port-Louis, Mauritius.

Brown, S. and Lugo, A.E. (1994) Rehabilitation of tropical lands: a key to sustaining development. *Restoration Ecology* 2(2), 97–111.

Buckley, P. (ed.) (1989) *Biological Habitat Reconstruction*. Belhaven Press, London.

Burch, F. (2001) Habitat restoration. In: Nash, K. and Burch, F. (eds) *Conservation of Biodiversity within Ecosystems*. Imperial College at Wye, University of London, London.

Cairns, Jr., J. (ed.) (1995) *Rehabilitating Damaged Ecosystems*. Lewis, New York.

Callenbach, E. (1995) *Bring Back the Buffalo*. Island Press, Covelo, California.

Cary, G., Lindenmayer, D. and Dovers, S. (eds) (2003) *Australia Burning*. CSIRO, Melbourne, Australia.

Chapman, C.A. and Chapman, L.J. (1996) Exotic tree plantations and the regeneration of natural forests in Kibale National Park, Uganda. *Biological Conservation* 76, 253–257.

Chester, P.I. and Raine, J.I. (1990) Mana Island revegetation: data from late holocene pollen analysis. In: Towns, D.R., Daugherty, C.H. and Atkinson, I.A.E. (eds) *Ecological*

Restoration of New Zealand Islands. Conservation Sciences Publication No. 2, Department of Conservation, Wellington, New Zealand, pp. 113–122.

Child, L. and Wade, M. (2000) *The Japanese Knotweed Manual*. Packard, Chichester, UK.

Conover, M. (2001) *Resolving Human–Wildlife Conflicts: The Science of Wildlife Damage Management*. Lewis/CRC Press, Boca Raton, Florida.

CORINE (1991) *CORINE Biotopes Manual. Habitats of the European Community. A Method to Identify and Describe Consistently Sites of Major Importance for Nature Conservation*, 3 Vols. Commission of the European Communities, 321 pp. Available at: http://biodiversity-chm.eea.europa.eu/information/document/F1088156525/F1125582140.

Cortner, J. and Moote, M.A. (1999) *The Politics of Ecosystem Management*. Island Press, Washington, DC.

Cox, G. (1997) *Conservation Biology: Concepts and Applications*. W.C. Brown, Dubuque, Iowa.

Cracknell, J. (1983) Soil and water conservation in the Windward Islands. *World Crops* 35(6), 218–220.

Cronk, Q.C.B. (1989) The past and present vegetation of St Helena. *Journal of Biogeography* 16, 47–64.

Davies, C.E. and Moss, D. (1998) *EUNIS Habitat Classification. Final Report to the European Topic Centre on Nature Conservation, European Environment Agency, with Further Revisions to Marine Habitats*. November 1998. 204 pp.

Davis, M.A. (2005) Invasion biology 1958–2004: the pursuit of science and conservation. In: Cadotte, M.W., McMahon, S.M. and Fukami, T. (eds) *Conceptual Ecology and Invasion Biology: Reciprocal Approaches to Nature*. Kluwer, London. Available at: http://www.cedarcreek.umn.edu/biblio/fulltext/t1972.pdf.

de Klemm, C. (1996) *Introductions of Non-native Organisms into the Natural Environment*. Nature and Environment Series No. 73. Council of Europe, Strasbourg, France.

Devilliers, P. and Devilliers-Terschuren, J. (1996) *A Classification of Palearctic Habitats*. Nature and Environment series 78. Council of Europe, Strasbourg, France.

Diggelen, R., Grootjans, A.P. and Harris, J.A. (2001) Ecological restoration: state of the art or state of the science. *Restoration Ecology* 9, 115–118.

Dodson, A.P., Bradshaw, A.D. and Baker, A.J.M. (1997) Hopes for the future: restoration ecology and conservation biology. *Science* 277, 515–522.

Donlan, C.J., Howald, G.R., Tershy, B.R. and Croll, D.A. (2003) Evaluating alternative rodenticides for island conservation: roof rat eradication from the San Jorge Islands, Mexico. *Biological Conservation* 114(1), 29–34.

Draper, D., Marques, I., Rosselló-Graell, A. and Iriondo, J.M. (2006a) Definition of a micro-reserve for endangered *Narcissus cavanillesii* in Portugal. In Aguilella, A., Ibars, A.M., Laguna, E. and Pérez, B. (eds) *Proceedings of the 4th European Conference on the Conservation of Wild Plants, Planta Europa*. Generalitat Valenciana and Universitat de Valencia, Valencia, Spain. Available at: http://www.nerium.net/plantaeuropa/Download/Procedings/Draper_et_al.pdf.

Draper, D., Rosselló-Graell, A., Marques, I. and Iriondo, J.M. (2006b) Translocation action of *Narcissus cavanillesii*: selecting and evaluating the receptor site. In: Aguilella, A., Ibars, A.M., Laguna, E. and Pérez, B. (eds) *Proceedings of the 4th European Conference on the Conservation of Wild Plants, Planta Europa*. Generalitat Valenciana and Universitat de Valencia, Valencia, Spain. Available at: http://www.nerium.net/plantaeuropa/Download/Procedings/Draper_Rossello_et_al.pdf.

Dulloo, M.E. and Maxted, N. (2007) The Crop Wild Relative Specialist Group of the IUCN Species Survival Commission. In: Maxted, N.,Ford-Lloyd, B.V., Kell, S.P., Iriondo, J.M., Dulloo, E. and Turok, J. (eds) *Crop Wild Relative Conservation and Use*. CAB International, Wallingford, UK.

Dulloo, M.E., Verburg, J., Paul, S.S., Green, S.E., Boucherville-Baissac, P. de and Jones, C. (1997) *Ile aux Aigrettes Management Plan 1997–2001*. Mauritian Wildlife Foundation Technical Series No. 1/97, Mauritius.

Dulloo, M.E., Kell, S.P. and Jones, C.G. (2002) Impact and control of invasive alien species on small islands. *International Forestry Review* 4(4), 277–285.

Dyer, R., Jacklyn, P., Partridge, I., Russell-Smith, J. and Williams, D. (eds) (2001) *Savanna Burning. Understanding and Using Fire in Northern Australia*. Tropical Savannas CRC, Darwin, Australia.

EEA (1993–2007) *European Nature Information System (EUNIS)*. European Environment Agency. Available at: http://eunis.eea.europa.eu/index.jsp.

El-Demerdash, M.A. (1996) The vegetation of the Farasãn Islands, Red Sea, Saudi Arabia. *Journal of Vegetation Science* 7, 81–88.

ENEBRO (no date) *Restoration of Coastal Dunes with* Juniperus *spp. in Valencia*. Available at: http://www.lifeenebro.com/LifeEnebro/ENG/Inicial.htm.

Eskildsen, L.I., Olesen, J.M. and Jones, C.G. (2004) Feeding response of the Aldabra giant tortoise (*Geochelone gigantea*) to island plants showing heterophylly. *Journal of Biogeography* 31(11), 1785.

European Commission (2007) *Interpretation Manual of European Union Habitats, EUR 27*. European Commission, Brussels, Belgium. Available at: http://ec.europa.eu/environment/nature/legislation/habitats directive/docs/2007_07_im.pdf

European Communities (1995–2007a) *Council Directive 92/43/EEC of 21 May 1992 on the Conservation of Natural Habitats and of Wild Fauna and Flora*. Available at: http://eur-lex.europa.eu/LexUriServ/LexUriServ.do?uri=CELEX:31992L0043:EN:HTML.

European Communities (1995–2007b) *Life-Nature Publications*. Available at: http://ec.europa.eu/environment/life/publications/lifepublications/index.

Falk, D.A. and Holsinger, K.E. (eds) (1991) *Genetics and Conservation of Rare Plants*. Oxford University Press, Oxford.

Falk, D.A., Millar, C.I. and Olwell, M. (1996) Guidelines for developing a rare plant reintroduction plan. In: Falk, D.A., Millar, C.I. and Olwell, M. (eds) *Restoring Diversity: Strategies for Reintroduction of Endangered Plants*. Island Press, Washington, DC, pp. 453–490.

Fernández-Orueta, J. and Aranda, Y. (2001) *Methods to Control and Eradicate Non-native Terrestrial Vertebrate Species*. Nature and Environment Series No. 118. Council of Europe, Strasbourg, France.

Fernández-Reguera, A. (2001) Seafront of the Albufera, Dehesa del Saler, sector. Valencia. *Via Arquitectura* 10, 76. Available at: http://www.via-arquitectura.net/10/10-076.htm.

Ferris, R. and Carter, C.I. (2000) *Managing Rides, Roadsides and Edge Habitats in Lowland Forests*. FC Bulletin No. 123. Forestry Commission, Edinburgh, UK.

Flenley, J.R. and King, S.M. (1984) Late quaternary pollen records from Easter Island. *Nature* 307, 47–50.

Ford-Lloyd, B.V., Kell, S.P. and Maxted, N. (2007) Establishing conservation priorities for crop wild relatives. In: Maxted, N., Ford-Lloyd, B.V., Kell, S.P., Iriondo, J.M., Dulloo, E. and Turok, J. (eds) *Crop Wild Relative Conservation and Use*. CAB International, Wallingford, UK.

Forestry Commission (1995) *Forests and Archaeology Guidelines*. Forestry Commission, Edinburgh, UK.

Frankel, O.H., Brown, A.H.D. and Burdon, J.J. (eds) (1995) *The Conservation of Plant Biodiversity*. Cambridge University Press, Cambridge.

FSC (2003) *Forest Stewardship Council Web site*. Available at: http://www.fsc.org/en/.

FWS (2005) *Endangered Species Habitat Conservation Planning*. US Fish and Wildlife Service, Washington, DC. Available at: http://www.fws.gov/endangered/hcp/index.html.

FWS (2007a) *Recovery and Delisting of Endangered Species*. US Fish and Wildlife Service. Available at: http://www.fws.gov/endangered/recovery/index.html.

FWS (2007b) *FWS Publications Online*. US Fish and Wildlife Service. Available at: http://library.fws.gov/pubs3.html.

FWS and NMFS (1996) *Habitat Conservation Planning Handbook*. US Fish and Wildlife Service and National Oceanic and Atmospheric Administration Fisheries Service, Washington, DC. Available at: http://www.fws.gov/endangered/hcp/hcpbook.html.

Gade, D.W. (1985) Man and nature on Rodrigues: tragedy of an island common. *Environmental Conservation* 12(3), 207–216.

García, F., Crawford, R.M.M. and Díaz, M.C. (eds) (1997) *The Ecology and Conservation of European Dunes*. University of Seville, Seville, Spain.

Genovesi, P. and Shine, C. (2003) *European Strategy on Invasive Alien Species*. Nature and Environment series No. 137. Council of Europe, Strasbourg, France. Available at: http://www.coe.int/t/e/cultural_co-operation/environ ment/nature_and_biological_diversity/nature_protection/sc23_tpvs07erev.pdf?L=E.

Gilbert, O.L. and Anderson, P. (1998) *Habitat Creation and Repair*. Oxford University Press, Oxford/New York.

Given, P. (1994) *Principles and Practice of Plant Conservation*. Timber Press, Portland, Oregon.

Grayson, C. (ed.) (2000) *Restoration of Lost Human Uses of the Environment*. SETAC Press, Pensacola, Florida.

Grove, R.H. (1995) *Green Imperialism: Colonial Expansion, Tropical Island Edens, and the Origins of Environmentalism*. Cambridge University Press, Cambridge.

Guerrant, Jr., E.O. and Pavlik, B.M. (1998) Reintroduction of rare plants: genetics, demography and the role of *ex situ* conservation methods. In: Fiedler, P.L. and Kareiva, P.M. (eds) *Conservation Biology: For the Coming Decade*. Chapman & Hall, New York, pp. 80–108.

Haagsma, B. and Reij, C. (1993) Frentes de Trabalho: Potentials and limitations of large-scale labour employment for soil and water conservation in Cape Verde. *Land Degradation and Rehabilitation* 4, 73–85.

Hambler, C. (2003) *Conservation*. Cambridge University Press, Cambridge.

Hammitt, W.E. and Cole, D.N. (1998) *Wildland Recreation. Ecology and Management*, 2nd edn. Wiley, New York.

Harker, D., Harker, K., Evans, M. and Evans, S. (1990) *USGA Landscape Restoration Handbook*. United States Golf Association/CRC Press, Boca Raton, Florida.

Harris, J.A., Birch, P. and Palmer, J. (1996) *Land Restoration and Reclamation. Principles and Practices*. Addison-Wesley Longman (Pearson Education), Essex, UK.

Harris, J.A., Hobbs, R.J., Higgs, E. and Aronson, J. (2006) Ecological restoration and global climate change. *Restoration Ecology* 14(2), 170–176.

Higgs, E. (2003) *Nature by Design. People, Natural Process, and Ecological Restoration*. MIT Press, Cambridge, Massachusetts.

Hill, M.O. (1994) *DECORANA and TWINSPAN, for Ordination and Classification of Multivariate Species Data*. Centre for Ecology and Hydrology, Huntingdon, UK. Available at: http://www.ceh.ac.uk/products/software/CEHSoftware-DECORANATWINSPAN.htm.

Hobbs, R.J. and Saunders, D.A. (1992) *Reintegrating Fragmented Landscapes*. Springer, New York.

Howard, G.H. (2000) *Control Options: Freshwater Invasives*. Paper presented to the Workshop on 'Best Management Practices for preventing and controlling Invasive Alien Species', Cape Town, South Africa, 22–24 February, 2000. Available at: http://www.iucn.org/webfiles/doc/archive/2001/IUCN853.doc.

Hughes, F.M.R. (2000) Popular tales from the riverbank. *River Restoration News* 7, 2. Available at: http://therrc.co.uk/newsletters/issue7.pdf.

Hussey, B.M.J., Hobbs, R.J. and Saunders, D.A. (1991) *Guidelines for Bush Corridors*. Surrey Beatty, Chipping Norton, New South Wales, Australia.

ISSG (2000) *Guidelines for the Prevention of Biodiversity Loss Caused by Alien Invasive Species*. Prepared by the IUCN/SSC Invasive Species Specialist Group and approved by the 51st Meeting of the IUCN Council. IUCN, Gland, Switzerland, February 2000. Available at: http://www.iucn.org/themes/ssc/publications/policy/invasivesEng.htm.

IUCN (1995–2006) *Slow but Steady – Giant Tortoises Help Clear Invading Species from Island Paradise*. International Union for Conservation of Nature and Natural Resources. Available at: http://www.iucn.org/themes/ssc/news/ile_aux_aigrettes.htm.

Plate 9

Plate 10

Plate 9 Restoration of dune vegetation
Restoration of dune vegetation in El Saler, Nature Park L'Albufera (Valencian Community, Spain). The dunes are divided into 5 x 5 m quadrats and protected from the wind with low fences made with local materials (e.g., dry leaves of *Ammophila arenaria*). Each quadrat is planted with a pre-selected combination of seeds, bulbs, rhizomes and plantets. As many as 120 species are used in this project, including endangered species, such as *Otanthus maritimus*. (Photo credit: Emilio Laguna)

Plate 10 Fencing to protect from herbivory
More than 300 fenced areas save several habitat types from the effects of the overpopulation of Spanish wild goats (*Capra hispanica*) in the Nature Park of Cazorla (Andalucia, Spain). This picture shows the recovery of karst vegetation (*Taxus baccata, Sorbus torminalis, Crataegus monogyna,* etc.) in an eight-year-old fenced area which provides a habitat for the endemic *Atropa baetica*. (Photo credit: Emilio Laguna)

Plate 11

Plate 12

Plate 11 Genebank
It is not possible to ensure the safe conservation of all the diversity within a given species *in situ*. An adequate representative sample of individual plants, specific genotypes or sub-populations from the wild needs to be conserved in a genebank, such as the one illustrated, to facilitate and promote their use and/or subsequent exchange with users. The genebank of the Central Research Institute for field crops in Ankara Turkey conserves many wild relatives of wheat as a complementary strategy for their conservation. (Photo credit: Ehsan Dulloo)

Plate 12 Native plant nursery
The habitats of wild populations are often degraded and require management interventions to enhance the target taxon population at the reserve site. The material required for population enhancement can either come directly from the wild or from *ex situ* sources. Here a temporary nursery has been built on the Ile aux Aigrettes Nature Reserve (Mauritius) to propagate native plants for use in a restoration programme. (Photo credit: Ehsan Dulloo)

IUCN (1998) *Guidelines for Re-introductions.* Prepared by the IUCN/SSC Re-introduction Specialist Group, IUCN, Gland, Switzerland/ Cambridge. Available at: http://www.iucnss crsg.org/downloads.html

Japanese Knotweed Alliance (no date) *Japanese Knotweed Alliance Web site*. Available at: http://www.cabi-bioscience.org/html/ japanese_knotweed_alliance.htm.

JNCC (2001–2006) *UK Biodiversity Action Plan Web site*. Joint Nature Conservation Committee. Available at: http://www.ukbap. org.uk/.

Johnson, N.C., Malk, A.J., Szaro, R.C. and Sexton, W.T. (eds) (1999) *Ecological Stewardship: A Common Reference for Ecosystem Management*. Elsevier, The Hague, The Netherlands.

Jones, T.A. (2003) The Restoration Gene Pool Concept: beyond the native versus non-native debate. *Restoration Ecology* 11(3), 281–290.

Jones, T.A. and Monaco, T.A. (2007) A restoration practitioner's guide to the Restoration Gene Pool Concept. *Ecological Restoration* 25(1), 12–19.

Jordan, W.R. (2003) *The Sunflower Forest. Ecological Restoration and the New Communion with Nature*. California University Press, Berkeley, California.

Jordan, W.R., Gilpin, M.E. and Aber, J.D. (1987) *Restoration Ecology*. Cambridge University Press, Cambridge.

Jordan, W.R., Gilpin, M.E. and Aber, J.D. (eds) (1990) *Restoration Ecology. A Synthetic Approach to Ecological Research.* Cambridge University Press, Cambridge.

Juan, A., Crespo, M.B., Cowan, R.S., Lexer, C. and Fay, M. (2004) Patterns of variability and gene flow in *Medicago citrina*, an endangered endemic of islands in the western Mediterranean, as revealed by amplified fragment length polymorphism (AFLP) *Molecular Ecology* 13(9), 2679–2690.

Kell, S.P. (1996) A vegetation survey of a potential genetic reserve on the Indian Ocean island of Rodrigues. MSc thesis, The University of Birmingham, Birmingham, UK.

Kell, S.P., Knüpffer, H., Jury, S.L., Maxted, N. and Ford-Lloyd, B.V. (2005) *Catalogue of Crop Wild Relatives for Europe and the Mediterranean*. University of Birmingham, Birmingham, UK. Available at: http://cwris. ecpgr.org/(CWRIS) and on CD-ROM.

Kell, S.P., Knüpffer, H., Jury, S.L., Ford-Lloyd, B.V. and Maxted, N. (2007a) Crops and wild relatives of the Euro-Mediterranean region: making and using a conservation catalogue. In: Maxted, N., Ford-Lloyd, B.V., Kell, S.P., Iriondo, J.M., Dulloo, E. and Turok, J. (eds) *Crop Wild Relative Conservation and Use*. CAB International, Wallingford, UK.

Kell, S.P., Moore, J.D., Iriondo, J.M., Scholten, M.A., Ford-Lloyd, B.V. and Maxted, N. (2007b) CWRIS: an information management system to aid crop wild relative conservation and sustainable use. In: Maxted, N., Ford-Lloyd, B.V., Kell, S.P., Iriondo, J.M., Dulloo, E. and Turok, J. (eds) *Crop Wild Relative Conservation and Use*. CAB International, Wallingford, UK.

Knapp, E.E. and Dyer, A.R. (1998) When do genetic considerations require special approaches to ecological restoration? In: Fiedler, P.L. and Kareiva, P.M. (eds) *Conservation Biology for the Coming Decade*. Chapman & Hall, New York, pp. 345–363.

Knight, R.L. and Landres, P.B. (eds) (1998) *Stewardship Across Boundaries*. Island Press, Washington, DC.

Krebs, C.J. (1989) *Ecological Methodology*. Harper & Row, New York.

Kuijken, E. (2003) *The Restoration of Sites and Ecological Corridors in the Framework of Building up the Pan-European Ecological Network, with Examples of Best Practices from European Countries*. Nature and Environment Series No. 135. Council of Europe, Strasbourg, France.

Kuusipalo, J., Ådjers, G., Jafarsidik, Y., Otsamo, A., Tuomela, K. and Vuokko, R. (1995) Restoration of natural vegetation in degraded *Imperata cylindrica* grassland: understory development in forest plantations. *Journal of Vegetation Science* 6, 205–210.

Laguna, E. (2001a) *The Micro-reserves as a Tool for Plant Conservation in Europe*. Nature and Environment Series No. 121. Council of Europe, Strasbourg, France.

Laguna, E. (2001b) The network of plant micro-reserves, a multifunctional instrument for awareness raising, involving landowners and scientific research. In: *Proceedings of the 2nd International Symposium of the Pan-Ecological Network: The Partnership of Local and Regional Authorities in the Conservation of Biological and Landscape Diversity (Rochefort, Belgium, 18–19 September 2000)*. Council of Europe, Strasbourg, France, pp. 99–103.

Laguna, E. (ed.) (2003) *Priority Habitats in the Valencian Community (Spain): Their Faunistic and Botanical Values/Hábitats Prioritarios de la Comunidad Valenciana*. CD-Rom edition, including English/Spanish PDF books. Conselleria de Territori i Habitatge, Generalitat Valenciana, Valencia, Spain.

Laguna, E. (2004) The plant micro-reserve initiative in the Valencian Community (Spain) and its use to conserve populations of crop wild relatives. *Crop Wild Relative* 2, 10–13. Available at: http://www.pgrforum.org/Documents/Newsletters/CWR_2_(online).pdf.

Laguna, E. (2005) Micro-reserves as a tool for grassland restoration and conservation in the Valencian Community (Spain). In: Struchkov, A. and Kuleshova, J. (eds) *Facets of Grassland Restoration*. The Open Country Series. Biodiversity Conservation Centre, Moscow, Russia, pp. 105–120.

Laguna, E. and Jiménez, J. (1995) Conservación de la flora de las islas Columbretes (España) *Ecologia Mediterranea* 21(1/2), 325–336.

Laguna, E., Ballester, G., Serra, L., Fabregat, C., Olivares, A., Deltoro, V., Pérez-Rovira, P., Pérez-Botella, J., Escribá, M.C., Ranz, J. and Mateache, P. (2002) Micro-reserves as a tool to conserve plants and vegetation in big cities. In: Nemec, J. (ed.) *Praga 2000 Natura Megapolis (International Conference on Nature Conservation in Big Cities): Proceedings of the Conference*. Czech Agency for Nature Conservation and Landscape Protection, Praga, The Czech Republic.

Laguna, E., Ballester, G., Olivares, A., Serra, L., Pérez-Rovira, P., Deltoro, V.I., Pérez-Botella, J. and Fos, S. (2003) Conservation of priority habitats in the Valencian Community, Spain (Project LIFE99 NAT/E/006417). *Ecologia Mediterranea* 29(1), 109.

Laguna, E., Ballester, G., Serra, L., Pérez-Rovira, P., Olivares, A., Deltoro, V.I., Pérez-Botella, J., Fos, S. and Fabregat, C. (2004) EU-funded project restores threatened plant habitats in Valencia, Spain. *PlantTalk* 35, 14.

Lambinon, J. (1997) *Introductions of Non-native Plants into the Natural Environment*. Nature and Environment Series No. 87. Council of Europe, Strasbourg, France.

Leck, M.A., Parker, V.T. and Simpson, R.L. (eds) (1989) *Ecology of Soil Seed Banks*. Academic Press, New York.

Leppakoski, E., Gollasch, S. and Olenin, S. (eds) (2002) *Invasive Aquatic Species of Europe. Distribution, Impacts and Management*. Kluwer, Dordrecht, The Netherlands.

Lill, D.O. (2000) *America's Vanishing Flora: Stories of Endangered Plants from the Fifty States and Efforts to Save Them*. Centre for Plant Conservation, St Louis, Missouri.

Lopes, V.L. and Meyer, J. (1993) Watershed management programme on Santiago Island, Cape Verde. *Environmental Management* 17(1), 51–57.

Luken, J.O. and Thieret, J.W. (1997) *Assessment and Management of Plant Invasions*. Springer, New York.

Lynch, M. (1991) The genetic interpretation of inbreeding depression and outbreeding depression. *Evolution* 45(3), 622–629.

Lynch, M. (1996) A quantitative genetic perspective on conservation issues. In: Avise, J.C. and Hamrick, J.L. (eds) *Conservation Genetics: Case Histories from Nature*. Chapman & Hall, New York. pp. 471–501.

Mack, R.N., Simberloff, D., Lonsdale, W.M., Evans, H., Clout, M. and Bazzaz, F.A. (2000) Biotic invasions: causes, epidemiology, global consequences, and control. *Ecological Applications* 10, 689–710.

Makhzoumi, J. and Pungetti, G. (1999) *Ecological Landscape Design and Planning, The Mediterranean Context*. Chapman & Hall/Spon-Routledge, London.

Maltby, E., Holdgate, M., Acreman, M. and Weir, A. (eds) (1999) *Ecosystem Management: Questions for Science and Society*. Royal Holloway Institute for Environmental Research, London

Marren, P. (1999) *Britain's Rare Flowers*. Academic Press, London.
Marrero-Gómez, M.V., Bañares-Baudet, A. and Carqué-Alamo, E. (2003) Plant resource conservation planning in protected natural areas: an example from the Canary Islands, Spain. *Biological Conservation* 113, 399–410.
Matthews, A. (2002) *Where the Buffalo Roam: Restoring America's Great Plains*. University of Chicago Press, Chicago, Illinois.
Maxted, N., Ford-Lloyd, B.V. and Hawkes, J.G. (1997) *Plant Genetic Conservation – the* In Situ *Approach*. Chapman & Hall, London.
McKay, J.K., Christian, C.E., Harrison, S. and Rice, K.J. (2005) How local is local? – a review of practical and conceptual issues in the genetics of restoration. *Restoration Ecology* 13(3), 432–440.
McNeely, J.A., Mooney, H.A., Neville, L.E., Schei, P.J. and Waage, J.K. (eds) (2001) *A Global Strategy on Invasive Alien Species*. IUCN, Gland, Switzerland/Cambridge.
Meffe, G.K., Nielsen, L.A., Knight, R.L. and Schenborn, D.A. (2002) *Ecosystem Management: Adaptative Community-based Conservation*. CSIRO, Melbourne, Australia.
Merton, D.V., Atkinson, I.A.E., Strahm, W., Jones, C., Empson, R.A., Mungroo, Y., Dulloo, E. and Lewis, R. (1989) *A Management Plan for the Restoration of Round Island, Mauritus*. Jersey Wildlife Preservation Trust/Ministry of Agriculture, Fisheries and Natural Resources, Mauritius.
Miller, C.J., Craig, J.L. and Mitchell, N.D. (1994) Ark 2020: a conservation vision for Rangitoto and Motutapu Islands. *Journal of the Royal Society of New Zealand* 24(1), 65–90.
Mitchell, R.S., Sheviak, C.J. and Leopold, D.J. (eds) (1990) *Ecosystem Management: Rare Species and Significant Habitats*. New York State Museum Bulletin, No. 471 (Proceedings of the 15th Annual Natural Areas Conference). New York State Museum Bulletin, New York.
Moffat, A.J. (1994) *Reclaiming Disturbed Land for Forestry*. Forestry Commission Bulletin No. 110. Forestry Commission, Edinburgh, UK
Montesinos, D., Garcia-fayos, P. and Mateu, I. (2006) Conflicting selective forces underlying seed dispersal in the endangered plant *Silene diclinis*. *International Journal of Plant Sciences* 167(1), 103–110.
Montmollin, B. de and Strahm, W. (eds) (2005) *The Top 50 Mediterranean Island Plants. Wild Plants at the Brink of Extinction and What Is Needed to Save Them*. IUCN/SSC Mediterranean Islands Plant Specialist Group, IUCN, Gland, Switzerland/Cambridge.
Mooney, H.A. and Hobbs, R.J. (eds) (2002) *Invasive Species in a Changing World*. Island Press, Washington, DC.
Mooney, H.A., Mack, R.N., McNeely, J.A., Neville, L.E., Schei, P.J. and Waage, J.K. (eds) (2005) *Invasive Alien Species. A New Synthesis*. SCOPE Series, No. 63. Shearwater Books, Washington, DC.
Morse, L.E. and Henifin, M.S. (1981) *Rare Plant Conservation. Geographic Data Organization*. New York Botanical Garden Press, New York.
Munshower, F.F. (1994) *Practical Handbook of Disturbed Land Revegetation*. CRC Press, Boca Raton, Florida.
Myers, J. and Bazely, D. (eds) (2003) *Ecology and Control of Introduced Plants*. Cambridge University Press, Cambridge.
OEC–DIREN (1998) *Bilan et Prospective 1994–1997: Conservation des Habitats Naturels et des Espèces Végétales d'interêt Communitaire Prioritaire de la Corse*. Office de l'Environnement de la Corse (OEC) and Direction Régionale de l'Environnement (DIREN), Corte, Corse.
O'Malley, P.G. (1991) Large-scale restoration on Santa Catalina Island, California. *Restoration and Management Notes* 9(1), 7–15.
Otten, A., Alphenaar, A., Pijls, Ch., Spuij, F. and de Wit, H. (1997) In situ *Soil Remediation*. Kluwer, Dordrecht, The Netherlands.
Oxford University Press (1993) *The New Shorter Oxford English Dictionary*. Oxford University Press, New York/Clarendon Press, Oxford.
Parker, D.M. (1995) *Habitat Creation. A Critical Guide*. English Nature Science Series No. 21. English Nature, Peterborough, UK.
Parrotta, J.A. (1995) Influence of overstory composition on understory colonization by native species in plantations on a degraded tropical site. *Journal of Vegetation Science* 6, 627–636.

PGR Forum (2005) *Crop Wild Relative Information System (CWRIS)*. University of Birmingham, UK. Available at: http://cwris.ecpgr.org/.

Perez-Bañon, C. and Marcos, M.A. (1998) Life history and description of the immature stages of *Eumerus purpurariae* (Diptera: Syrphidae) developing in *Opuntia maxima*. *European Journal of Entomology* 95(3), 373–382. Available at: http://www.eje.cz/pdfarticles/416/eje_095_3_373_Perez.pdf.

Perrow, M.R. and Davy, A.J. (eds) (2002) *Handbook of Ecological Restoration*. 2 Volumes. Cambridge University Press, Cambridge.

Pirot, J.Y., Meynell, P.J. and Elder, D. (eds) (2000) *Ecosystems Management: Lessons from Around the World. A Guide for Development and Conservation Practitioners*. IUCN, Gland, Switzerland/Cambridge.

Primack, R. (1998) Monitoring rare plants. Keeping track of population changes over time is an important part of rare plant management. *Plant Talk* 15, 29–32.

Pyne, S.J. (1996) *Introduction to Wildland Fire: Fire Management in the United States*, 2nd edn. Wiley, Chichester, UK/New Jersey.

Pyusek, P., Prach, K., Rejmanek, M. and Wade, M. (1995) *Plant Invasions: General Aspects and Special Problems*. SPB, Amsterdam, The Netherlands.

Pywell, R.F., Cox, R., Pakeman, R.J., Stevenson, M., Roy, D. and Ashton, C. (1996) *A Decision Support System for Habitat Restoration: First Report*. Report to the Ministry of Agriculture, Fisheries and Food – July 1996, London.

Rana, B.C. (ed.) (1998) *Damaged Ecosystems and Restoration*. World Scientific, River Edge, New Jersey.

Ray, G.J. and Brown, B.J. (1995) Restoring Caribbean dry forests: evaluation of tree propagation techniques. *Restoration Ecology* 3(2), 86–94.

Rieseberg, L.H. and Swensen, S.M. (1996) Conservation genetics of endangered island plants. In: Avise, J.C. and Hamrick, J.L. (eds) *Conservation Genetics: Case Histories from Nature*. Chapman & Hall, New York, pp. 305–334.

Robichaux, R.H., Friar, E.A. and Mount, D.W. (1997) Molecular genetic consequences of a population bottleneck associated with the reintroduction of the Mauna Kea Silversword (*Argyroxiphium sandwicense* ssp. *sandwicense* [Asteraceae]). *Conservation Biology* 11(5), 1140–1146.

Rohlf, F.J. (1998) *Numerical Taxonomy System (NTSYSpc)*. Exeter Software, New York. Available at: http://www.exetersoftware.com/cat/ntsyspc/ntsyspc.html.

Rosselló-Graell, A., Draper, D., Correia, A.I.D. and Iriondo, J.M. (2002a) Translocación de una población de *Narcissus cavanillesii* A. Barra & G. López en Portugal como medida de minimización de impacto. *Ecosistemas* Año 11, No. 3. Available at: http://www.revistaecosistemas.net/pdfs/257.pdf.

Rosselló-Graell, A., Draper, D., Correia, A.I.D. and Iriondo, J.M. (2002b) Conservation program for *Narcissus cavanillesii* A. Barra & G. López (Amaryllidaceae) in Portugal. A translocation action. *OPTIMA Newsletter* 36, 18–19.

RSG (no date) IUCN/SSC Re-introduction Specialist Group (RSG) Web site. Available at: http://www.iucnsscrsg.org/rsghome.html.

Ruiz-Jaen, M.C. and Aide, T.M. (2005) Restoration success: how is it being measured? *Restoration Ecology* 13(3), 569–577.

Russell-Smith, J., Ryan, P.G. and Cheal, D. (2002) Fire regimes and the conservation of sandstone heath in monsoonal northern Australia: frequency, interval, patchiness. *Biological Conservation* 104, 91–106.

Salafsky, N., Margoluis, R. and Redford, K. (2001) *Adaptive Management: A Tool for Conservation Practitioners*. Biodiversity Support Programme, Washington, DC. Available at: http://fosonline.org/resources/Publications/AdapManHTML/Adman_1.html.

Samson, F.B. and Knopf Fritz, L. (eds) (1996) *Ecosystem Management. Selected Readings*. Springer, Berlin, Germany/Heidelberg, Germany/New York.

Sandlund, O.T., Schei, P.J. and Viken, A. (eds) (1999) *Invasive Species and Biodiversity Management*. Population and Community Biology Series No. 24, Kluwer, Dordrecht, The Netherlands.

Sanjaume, E. (1988) The dunes of El Saler. *Journal of Coastal Research* special issue 3, 63–70.

Sanjaume, E. and Pardo, J. (1991) The possible influence of sea-level rise on the precarious

dunes of Devesa del Saler Beach, Valencia, Spain. *Landscape Ecology* 6(1–2), 57–64.
Saunders, D., Hobbs, R.J. and Ehrlich, P.R. (1993) *Repairing a Damaged World: An Outline for Ecological Restoration*. Surrey Beatty, Chipping Norton, New South Wales, Australia.
SCB (2005) *Society for Conservation Biology Web site*. Available at: http://www.conbio.org/.
Schemnitz, S.D. (ed.) (1980) *Wildlife Management Techniques Manual*, 4th edn. The Wildlife Society, Washington, DC.
Schwartz, M.W. (1997) *Conservation in Highly Fragmented Landscapes*. Chapman & Hall, New York.
SER (no date) *Society for Ecological Restoration Online*. Available at: http://www.ser.org/default.asp.
SER International Science and Policy Working Group (2004) *The SER International Primer on Ecological Restoration*. Available at: www.ser.org and Society for Ecological Restoration International, Tucson, Arizona.
Shine, C. (1996) *Private or Voluntary Systems of Natural Habitats' Protection and Management*. Nature and Environment Series No. 85. Council of Europe, Strasbourg, France.
Shine, C., Williams, N. and Gundling, L. (2000) *A Guide to Designing Legal and Institutional Frameworks on Alien Invasive Species*. Environmental Policy and Law Papers, No. 40. IUCN, Gland, Switzerland/Cambridge/Bonn, Germany. Available at: http://www.iucn.org/dbtw-wpd/edocs/EPLP-040-En.pdf.
Simberloff, D. (1990) Reconstructing the ambiguous: can island ecosystems be restored? In: Towns, D.R., Daugherty, C.H. and Atkinson, I.A.E. (eds) *Ecological Restoration of New Zealand Islands*. Conservation Sciences Publication No. 2. Department of Conservation, Wellington, New Zealand, pp. 37–51.
Sivasithamparam, K., Dixon, K.W. and Barrett, R.L. (eds) (2002) *Microorganisms in Plant Conservation and Biodiversity*. Kluwer, Dordrecht, The Netherlands.
Smart, J., Imboden, Ch., Harper, M. and Radford, E. (eds) (2002) *Saving the Plants of Europe. European Plant Conservation Strategy*. Planta Europa and The Council of Europe, London.
Strahm, W. (1989) *Plant Red Data Book for Rodrigues*. IUCN, WWF.
Struchov, A. and Kuleshova, J. (eds) (2005) *Facets of Grassland Restoration*. The Open Country Series. Biodiversity Conservation Centre, Moscow, Russia.
Sutherland, W.J. and Hill, D.A. (eds) (1995) *Managing Habitats for Conservation*. Cambridge University Press, Cambridge.
Taylor, P. (1995) Whole ecosystem restoration: recreating wilderness. *ECOS* 16(2), 22–28.
Torres, E., Iriondo, J.M., Escudero, A. and Pérez, C. (2003) Analysis of within-population spatial genetic structure in *Antirrhinum microphyllum* (Scrophulariaceae). *American Journal of Botany* 90, 1688–1695.
Urbanska, K.M., Webb, N. and Edwards, P.J. (1999) *Restoration Ecology and Sustainable Development*. Cambridge University Press, Cambridge.
Vallee, L., Hogbin, T., Monks, L., Makinson, B., Matthes, M. and Rossetto, M. (2004) *Guidelines for the Translocation of Threatened Plants in Australia*, 2nd edn. Australian Network for Plant Conservation, Canberra, Australia.
Veitch, C.R. and Clout, M.N. (eds) (2002) *Turning the Tide: The Eradication of Invasive Species*. Proceedings of the International Conference on Eradication of Island Invasives. Occasional Papers Series. IUCN Species Survival Commission, Gland, Switzerland/Cambridge. Available at: http://www.hear.org/articles/turningthetide/turningthetide.pdf.
Verhoeven, J.T.A. (2001) Ecosystem restoration for plant diversity conservation. *Ecological Engineering* 17, 1–2.
de Waal, L., Child, L., Wade, M. and Brock, J. (eds) (1994) *Ecology and Management of Invasive Riverside Plants*. Landscape and Ecology Series. Wiley, Chichester, UK.
Walker, L.R. and del Moral, R. (2003) *Primary Succession and Ecosystem Rehabilitation*. Cambridge University Press, Cambridge.
Warren, A. and French, J. (eds) (2000) *Habitat Conservation. Managing the Physical Environment*. Wiley, Chichester, UK.
Warren, S.D. and Aschmann, S.G. (1993) Revegetation strategies for Kaho'olawe Island, Hawaii. *Journal of Range Management* 46(5), 462–466.

Watt, T.A. and Buckley, G.P. (1994) *Hedgerow Management and Nature Conservation*. Wye College Press, Kent, UK.

Werner, P. (1987) Reflections on mechanistic experiments in ecological restoration. In: Jordan, W.R., Gilpin, M.E. and Aber, J.D. (eds) *Restoration Ecology: A Synthetic Approach to Ecological Research*. Cambridge University Press, Cambridge, pp. 321–328.

Whisenant, S.G. (1999) *Repairing Damaged Wildlands: A Process-orientated, Landscape-scale Approach*. Cambridge University Press, Cambridge.

White, G. and Gilbert, J. (eds) (2003) *Habitat Creation Handbook for the Minerals Industry*. Royal Society for the Protection of Birds, Sandy, UK.

Williams, P.A. (1997) *Ecology and Management of Invasive Weeds*. Science and Research Series No. 7 Department of Conservation, Wellington, New Zealand.

Williamson, M. (1996) *Biological Invasions*. Chapman & Hall, London.

Winfield, M. and Hughes, F.M.R. (2002) Variation in *Populus nigra* clones: implications for river restoration projects in the United Kingdom. *Wetlands* 22(1), 33–48.

Wingate, D.B. (1985) *The Restoration of Nonsuch Island as a Living Museum of Bermuda's Pre-colonial Terrestrial Biome*. ICBP Technical Publication No. 3.

Wittenberg, R. and Cock, M.J.W. (eds) (2001) *Invasive Alien Species: A Toolkit of Best Prevention and Management Practices*. CAB International, Wallingford, UK.

World Bank (1995) *Mauritius: Biodiversity Restoration Project*. GEF project document. Report No. 14527-MAS.

Wyse-Jackson, P. and Akeroyd, J. (1994) *Guidelines to be Followed in the Design of Plant Conservation or Recovery Plans*. Nature and Environment Series No. 68. Council of Europe, Strasbourg, France.

Young, T.P. (2000) Restoration biology. *Biological Conservation* 92, 73–83.

Zdzislaw, P. and Belousova, I.P. (eds) (2004) *European Bison (*Bison bonasus*): Current State of the Species and Strategy for its Conservation*. Nature and Environment Series No. 141. Council of Europe, Strasbourg, France.

6 Complementing *In Situ* Conservation with *Ex Situ* Measures

J.M.M. Engels,[1] L. Maggioni,[1] N. Maxted[2] and M.E. Dulloo[1]

[1]*Bioversity International, Rome, Italy;* [2]*School of Biosciences, University of Birmingham, Edgbaston, Birmingham, UK*

6.1 Introduction

Conserving plant genetic resources *in situ,* i.e. where the material has evolved its distinctive characteristics, has an important advantage over an *ex situ* conservation approach in that the material of the target species is exposed to the natural environment and can continue to evolve. However, continued exposure to a dynamic environment can also have some downsides. For example, the target taxon (i.e. a crop wild relative (CWR) genotype, population or even the entire species) can be threatened by micro- and macro-ecological events, such as floods, droughts, agricultural and landscape developments or climatic changes, by biotic stress such as pests and diseases or by human interventions. Human interventions or interferences in the biological processes of the target taxon both at the micro- or macro-scale can have the most devastating effects, e.g. overexploitation of a species/population through the gathering of the entire plant, fruits or other parts that affect either

 Conserving Plant Genetic Diversity in Protected Areas (eds J.M. Iriondo, N. Maxted and M.E. Dulloo)

the generative reproduction of the target taxon or that lead directly to a critical reduction of the number of individuals. The management plan and the procedures put into practice as part of the establishment of a genetic reserve should, in principle, assess the risk of human interference at the level of genetic diversity of the target taxa and provide strategies to mitigate the most adverse impacts. However, the effectiveness of these measures needs to be constantly verified. Consequently, monitoring the genetic parameters of the conserved taxa, together with the relevant ecological and demographic parameters, is an important aspect of the conservation effort (see Iriondo *et al.*, Chapter 4, this volume). The availability of good baseline data will greatly facilitate informed decision-making with regard to specific conservation interventions when any of the agreed critical thresholds that relate to the threat of a given taxon are passed. *Ex situ* conservation will need to be primarily considered when the level of genetic diversity of the target taxa is recognized to be under such a threat that it cannot be easily reversed by reserve management measures (i.e. low number of remaining individuals or heavy risk of loss of genetic diversity due to uncontrollable factors, such as climatic change).

Ex situ conservation can also be used to conserve an adequate representative sample of individual plants, specific genotypes or subpopulations from the reserve in a genebank, together with the related information, to facilitate access (for breeding, research, training and/or education) and to promote their use. In cases of overexploitation of a given taxon in the wild, its domestication and cultivation by farmers has also been practised in a number of cases. The introduction into cultivation of proteas in South Africa is an example of this approach. The establishment and subsequent maintenance of an adequate *ex situ* collection rich in genetic diversity is the starting point for such an intervention and the most appropriate site for this type of collecting might be a diversity-rich genetic reserve. Another factor to keep in mind is the long-term stability of genetic reserves' arrangements which is linked to human decisions and orientations that may vary over time. Therefore, several reasons lead us to consider the benefits of adopting complementary (*in situ* and *ex situ*) strategies for conservation. Whenever possible, decisions on when to back up genetic reserve CWR material in *ex situ* conditions should be made on the basis of security and cost–benefit considerations as well as common sense.

6.2 *Ex Situ* Conservation Approaches

Historically, the formal conservation of crop genetic resources was almost by default *ex situ*. This is because crop genetic resources were largely used by plant breeders, who traditionally maintained their own breeding collections *ex situ* and it was these collections that formed the basis for many of the global and regional crop germplasm collections. The establishment of the International Board for Plant Genetic Resources (IBPGR) (now Bioversity International) in 1974 was largely based on the need to coordinate worldwide collecting efforts of threatened local varieties and landraces due to the success of the distribution of high-yielding varieties for many of the major food crops. Besides this coordinating role, IBPGR was also concerned with the development of methodologies and technologies for long-term conservation and initially these were all related to *ex situ* approaches. It was only in the late 1980s and early 1990s that the *in situ* (genetic reserve and on-farm) conservation of crop genetic resources entered conservation discussions, triggered by the Convention on Biological Diversity (CBD) that had (and possibly still has) a very strong focus on the *in situ* conservation of wild resources.

The first step of any *ex situ* conservation effort is the location of the material to be included in the genebank for its conservation and subsequent use, through collecting either in nature, the farmers' fields, markets, etc. In the context of this publication the focus is on collecting germplasm from nature, either inside or outside a protected area.

With the understanding that we are dealing with 'wild' genetic resources that occur in undisturbed habitats or in known 'disturbed' habitats, i.e. managed or the result of managed habitats, the reader is referred to Guarino *et al.* (1995), which provides a comprehensive, practical guide to plant exploration and *ex situ* collecting and includes population genetic theory, numerous examples of collecting different types of material, etc.

Although dried seed conservation in genebanks at sub-zero temperatures (−18°C is recommended for long-term conservation) remains the most widely employed form of *ex situ* conservation (Plate 11), other techniques are increasingly used. A brief description of the various *ex situ* techniques available that might be specifically linked to *in situ* conservation is provided in Box 6.1 with key references to their application. There is an extensive body of knowledge on this subject and reference is made to a few general publications (e.g. Engels and Wood, 1999; Hawkes *et al.*, 2000; Engels and Visser, 2003; Thormann *et al.*, 2006).

6.3 Why Is an Effective *In Situ/Ex Situ* Link Critical?

The justifications for the promotion of the *in situ/ex situ* linkage can be broadly grouped under the following headings: *ex situ* material as a source for *in situ* population enhancement, *ex situ* safety duplication, complementary conservation and genetic diversity utilization.

6.3.1 *Ex situ* material as a source for *in situ* population enhancement

It would ideally be hoped that genetic reserves are situated in the most appropriate locations, i.e. hot spots of genetic diversity, and sites with long-term management security and/or where there is already active conservation management which does not require many interventions. However, too often this ideal is not the reality. In many cases, the habitats of wild populations are degraded to varying extents and require management interventions to enhance the target taxon population at the reserve site.

The material required for population enhancement (see Kell *et al.*, Chapter 5, this volume) can either come directly from the wild or from *ex situ* sources. The introduction of material from *ex situ* sources (genebanks, nurseries, etc.) would need to be done with caution because the introduction of any alien germplasm into a protected area is likely to impact on the genetic integrity of the original population

Box 6.1. Summary of *ex situ* conservation techniques.

Seed storage

Seed storage in cold rooms or in freezers (Plate 11) is the most widely used technique for conserving plant genetic resources which have so-called orthodox seeds that can tolerate drying and exposure to sub-zero temperatures. It involves drying of the seeds, usually to 3–7% seed moisture content depending on the species, and storage at an appropriate temperature (−18°C is recommended for long-term storage). During storage the viability of seeds needs to be monitored at regular intervals and the accessions may need to be regenerated when the viability falls below established thresholds in order to maintain their genetic integrity.

Useful reference: Rao *et al.* (2006).

Field genebank

In field genebanks, plant genetic resources are kept as plants out in the field or in glasshouses. They are predominantly used for the conservation of clonally propagated crops (root, tuber and bulb crops), species that produce so-called recalcitrant seed, i.e. those that do not tolerate drying and exposure to low temperatures (e.g. cacao, rubber, oil palm, coffee and coconut) and species that rarely produce seed. Although the maintenance of germplasm in field genebanks is expensive and not very secure from a number of unavoidable hazards, it is the only practicable method for many *ex situ* conservation programmes.

Useful references: Thormann *et al.* (2006); Reed *et al.* (2004).

***In vitro* storage**

This conservation method involves the maintenance of explants in a sterile, pathogen-free environment with a synthetic nutrient medium and is widely used for the conservation of species which produce recalcitrant seeds, no seeds at all or that are commercially vegetatively propagated. Different *in vitro* conservation methods are available: (i) slow growth conservation by limiting the environmental conditions and/or the culture medium; an example of this is the temperature reduction technique (varying from 0°C to 5°C for cold-tolerant species, and 9–18°C for tropical species) which can be combined with a decrease in light intensity or storage in the dark; and (ii) synthetic seed technique, which aims to use somatic embryos as true seeds by encapsulating embryos in alginate gel, and can then be stored after partial dehydration and sown directly *in vivo*.

Useful references: Engelmann (1997); Janick *et al.* (1993).

Cryopreservation

Long-term storage is practicable through storage of seeds or *in vitro* cultures at ultra-low temperature, usually in liquid nitrogen (−196°C) as cryopreserved material. At this temperature all cellular divisions and metabolic processes are stopped and, consequently, plant material can be stored without alteration or modification for theoretically unlimited period. Protocols of this technique are rather species-specific and even genotype-specific and the major costs are involved in the preparatory steps placing the material in the cryo-tank.

Useful references: Engelmann (1997); Reed *et al.* (2004).

Botanic gardens and arboreta

Botanic gardens and arboreta also maintain a wide diversity of plant species, especially of wild plants, and include medical, aromatic, ornamental and plants of major socio-economic importance. The accessions are usually displayed in a garden or field and typically consist of one or a few individuals per accession and/or accessions per species. However, many botanic gardens also have seed storage facilities as well as tissue culture facilities for conserving more genetic diversity.

Useful references: Heywood (1991); Laliberté (1997).

Continued

Box 6.1. *Continued*

Pollen storage

Like seed, pollen can be dried (about 5% moisture content on a dry weight basis) and stored below 0°C. However, pollen has a relatively short life compared to seeds, although this varies significantly among species. It is understood that plants can only be obtained by fertilizing a seed-born plant or by tissue culture.

Useful references: Hoekstra (1995); Engels *et al.* (2007).

DNA storage

The DNA material can now be extracted and stored at −20°C in an alcoholic solution in Eppendorf tubes. This form of conservation is becoming more and more important as molecular techniques are evolving and the number of users of DNA increases. It is also understood that no plants can be regenerated from a DNA sample and that this method should be part of a complementary conservation approach.

Useful references: Adams (1997); de Vicente (2006).

and may even lead to outbreeding depression if hybrids between the native and alien populations are less fit for local environmental conditions than native populations. Therefore, it is important to ensure that the *ex situ* conserved germplasm for reintroduction was originally collected from that site or is genetically close to the host population. For example, in the Global Environment Facility (GEF)-funded project 'Conservation and Sustainable Use of Dryland Agro-Biodiversity in Jordan, Lebanon, Palestine Authority and Syria' (http://www.icarda.org/Gef.html), the target taxon populations were enriched at the reserve site using native *ex situ* germplasm collected at an earlier time or locally sourced germplasm of target taxa and transferred to the reserve site, either as whole plants or seeds. Similarly, on Ile aux Aigrettes Nature Reserve in Mauritius, a temporary nursery was built to propagate seedlings to be used in the restoration of the islet (Plate 12).

Ex situ germplasm material can also be used to reintroduce a target taxon that has become extinct at a site where it was previously known to exist. When genetic reserve populations must be enriched with alien genetic diversity (i.e. no local provenance material is available), *ex situ* germplasm should only be used if it meets three basic criteria: (i) the taxonomic identification of the germplasm has been verified; (ii) the material is sourced from a provenance that is relatively geographically close to the target site; and (iii) the material is sourced from populations that have a homoclinal and ecogeographic match with the populations being enriched. Historically, little attention has been paid to the genetic implications of the introduction of alien germplasm, but as stressed in Chapter 5, the closer the introduced alien germplasm is to the native, in terms of a genetic and homoclinic match, the more likely it is that the reintroduction or recovery will be successful.

6.3.2 *Ex situ* safety duplication

The need to preserve a safety duplicate of *in situ* conserved germplasm in genebanks is greater now than ever before, not only because of the impact of climate change on natural plant diversity (Thuiller *et al.*, 2005; Van Vuuren *et al.*, 2006) but also because of other factors and human interferences that threaten the survival of genetic material in a reserve. Human-induced climatic changes have accelerated global warming over the last 30 years (Osborn and Briffa, 2005). Temperature increases are predicted to be in the range of 1.1–6.4°C by 2100 (IPCC, 2007), which is likely to result in large-scale extinctions (Thomas *et al.*, 2004). Although genetic reserve management should help prevent some of the threats to the survival of target populations, unexpected fires, flooding, plagues, vandalism, etc. may affect the *in situ* conservation of genetic diversity. Moreover, variable selection pressure is expected to influence the number of species, population sizes and gene frequencies in a genetic reserve at any point in time. In this context, it is relatively easy to sample and collect a representative sample of genetic diversity for *ex situ* storage (Plate 11) while a reserve is being established. Then, if for some reason the original population declines or goes extinct, the manager can always obtain seeds from the genebank to attempt to restore the natural population using samples of the original population. Seeds collected from populations at the site are likely to stand the best chance of population re-establishment because they are genetically adapted to grow at the site, unless the loss was due to climate change.

6.3.3 Complementary conservation

It is now widely accepted that the use of a single conservation technique to conserve biodiversity places that diversity at risk, i.e. extreme weather conditions may cause the extinction of target populations in genetic reserves, prolonged power cut may place germplasm conserved in genebanks at risk or civil strife is likely to impact negatively on populations whether conserved *in situ* or *ex situ*. As such, complementary or integrated conservation involving a combination of both *in situ* and *ex situ* techniques, each with their advantages and disadvantages, is most likely to secure diversity for future use (Maxted *et al.*, 1997). Bioversity International defines complementary conservation as ‘the combination of different conservation actions, which together lead to an optimum sustainable use of genetic diversity existing in a target genepool, in the present and future’ (Dulloo *et al.*, 2005).

In making such conservation choices it is important to take a holistic view of the objectives of the conservation of target material and to place them in a wider context, whenever possible, as part of a development process. This will include both local planning and development, and national considerations on how to best go about conservation and use at the national level. In many countries, specific programmes for the conservation of plant genetic resources for food and agriculture have been established with the national genebank as one of the more visible components. It can be argued that the level at which the conserved material is made available could impact possible sources of financial support for long-term conservation. Besides these technical and socio-economic aspects, it is also important to carefully examine the availability of infrastructural and human resources, as well as the administrative and political environment of the conservation effort to avoid later constraints. The latter two aspects have a more direct impact on the sustainability of the conservation efforts and/or on the practicality of making the conserved material available to its users according to the existing institutional and national policies.

Choice between conservation methods may be dictated by the biology of the species. For instance, if the wild species does not form botanical seeds (as for apomictic species such as *Allium*, *Narcissus* and *Poa*), or can be vegetatively propagated (e.g. bananas, pineapples, fruit trees), the choice includes on-farm conservation (e.g. as part of field borders or hedges between fields), maintenance in field genebanks, and *in vitro* slow growth and/or cryopreservation (Sharrock and Engels, 1997). Each conservation technique has its specific advantages and disadvantages, and these will vary from species to species, country to country and possibly even over time. By combining two or more *ex situ* methods, we may be able to increase the security of the conservation effort (e.g. *in vitro* conservation versus field genebank maintenance; conserving cryopreserved tissues versus maintaining them under slow growth), increase coverage of sampled genetic diversity (e.g. *in situ* and on-farm conservation of wild and/or cultivated material versus their conservation *ex situ*), reduce the costs of conservation (e.g. *in situ* conservation of CWR tends to be cheaper than their *ex situ* conservation) and increase physical accessibility (combining *in vitro* slow growth with cryopreservation allows plantlets to be provided).

6.3.4 Genetic diversity utilization

Germplasm users are always likely to find easier access to germplasm via a genebank (Plate 11), which routinely deals with potential user enquiries, than through a genetic reserve or on-farm conservation project manager. The seasonality of the availability of seeds (the most common form of germplasm dispatched for use) when genetic diversity is conserved *in situ* means that seeds are only available for relatively short periods of the year, whereas seeds from genebanks are available throughout the year. Therefore, when the *in situ* germplasm from the reserve or on-farm project is duplicated and made available via a genebank, the genebank may be seen to act as a staging post for those wishing to utilize the germplasm originally conserved *in situ*.

However, one of the major limitations to germplasm use is lack of characterization or evaluation data – how can users decide which germplasm to use if there is no way of distinguishing which is most fit for their purposes? This limitation may be thought to be compounded in the case of *in situ* conserved genetic diversity, but if the reserve manager wishes to ensure sustainability of conservation, use is critical to success. Possibly because of the potential magnitude of the task of *in situ* characterization or because protected area managers do not routinely undertake characterization or evaluation trials, it seems unlikely that actual *in situ* characterization or evaluation is feasible without significant additional resources. This in turn seems unlikely when resources are so limited for *ex situ* characterization. However, one way of circumventing this problem is to undertake 'virtual' or 'predictive' characterization.

Virtual or predictive characterization is the remote characterization of germplasm based on the ecological conditions under which the natural populations exist, using population passport data that are much more readily available. Simply knowing the provenance of CWR populations means a significant amount of characterization data can be deduced using contemporary geographic information system (GIS) techniques. For example, if a plant breeder works in a country that is likely to suffer lower rainfall following climate change, he or she can search for adaptation to drought by sourcing germplasm from slightly drier conditions that currently occur in the breeder's target country. Also, the breeder might look for germplasm from areas that are suffering other environmental stresses, e.g. disease hot spots, as germplasm from these areas may have adapted to local conditions and these useful traits may prove important to further develop cultivated plants. GIS analysis to characterize populations, by overlaying distribution maps of CWR with GIS-layers on environmental data, including climate and soil or pest and disease occurrence, is becoming routine. Furthermore, statistical analysis techniques can be used to classify and predict the distribution of certain characteristics. Although this form of 'virtual' or 'predictive' *in situ* characterization is likely to be conducted centrally, rather than within individual reserves or on-farm projects, the process should greatly enhance the use of *in situ* conserved germplasm and, therefore, conservation security.

6.4 Practical Considerations and Decision Making as Part of Management Procedures

Conserving genetic resources *ex situ* does not necessarily solve all the problems that one might encounter when conserving target taxa *in situ* in a genetic reserve. When collecting material from the genetic reserve for genebank storage, an important question to answer is what the actual unit of conservation is, i.e. a natural population, a subpopulation or a group of individuals. Furthermore, it is important to know what the biology of the species is, i.e. annual or perennial, inbreeding or outbreeding, herbaceous or woody, etc. The answers to such questions will determine how to best sample the target population, which *ex situ* conservation method or methods to use (see complementary conservation, Section 6.3.3) and how to best prepare the material for long-term storage.

When a given wild species produces orthodox seeds, it would be easiest to store the dried seeds in hermetically closed and vapour-proof containers at low temperature for the long term. The target population/individuals have to be adequately sampled in order to capture the existing genetic diversity with an agreed probability level. Furthermore, a sufficient amount should be collected to have an adequate stock for research purposes and distribution to possible users, as well as for viability testing and representative duplicates. In the case of an outbreeding species, the accession will have to be managed as one or more samples of the same population, all containing the same genetic diversity. In the case of an inbreeding species, one could consider splitting the original sample into accessions, each consisting of a pure line. Thus, later management of the material would be facilitated without fearing genetic drift/shift as well as the use of the material, especially when molecular tools are applied and/or the accession is evaluated for specific traits/characteristics. When specific traits/genotypes have been identified in an *ex situ*-conserved population, the only practical solution is to strive towards the maintenance of the genetic integrity of the originally collected sample and to incorporate the different genotypes as separate accessions in the collection. Details on accession-level management of germplasm can be found in Sackville Hamilton *et al.* (2002).

If the target material cannot be stored as seed, e.g. recalcitrant seed-producing species, the conservation procedures will be less simple. The sampling of the target material will be guided by the maximum number of individuals that can be conserved, either as entire plants in a field genebank or as tissue in an *in vitro* collection and/or cryopreserved. The storage of pollen could also be considered as an additional back-up and/or means to facilitate distribution, depending on the presence of quarantine pests and diseases.

One immediate measure whenever a target taxon is at risk of extinction is to include samples of that target taxon in one or more *ex situ* genebanks. A similar action will be required when the level of genetic diversity of the target taxon is shown to be reducing, as a result of the monitoring exercise. Collecting representative samples at regular intervals could also be considered, should the monitoring exercise reveal significant changes in the type of existing genetic diversity (i.e. in gene frequencies). The actual level of backing-up target taxa conserved in a genetic reserve in genebanks should be directly dependent on the results of monitoring and, therefore, decision making will be linked to the specific conservation objectives and defined parameters for intervention. Obviously, systematic monitoring at the genetic level of one or more taxa can be very expensive and time-consuming, and this will likely be possible only in a few selected cases. One more practical, less expensive alternative in the case of seed-propagated species is to collect representative population samples at least once for each target taxon in the reserve, and possibly complement this with additional samples collected at 10-year intervals (or more, depending on the level of observed changes in the reserve). In this case, the back-up of the evolving genetic diversity is secured on probabilistic assumptions and may be economically feasible.

6.5 Improving the *In Situ/Ex Situ* Conservation and Utilization Link

As noted throughout this text, there is an explicit link between the expenditure of resources on conservation and utilization of conserved material for human benefit. We must attempt to ensure that we make maximum use of *in situ* and *ex situ* conserved germplasm. Simmonds (1962) graphically suggested that mismanaged or underutilized germplasm collections may be regarded as mere museum exhibits gathering dust and by extension are unworthy of sustainable funding. This could be equally applied to those managing *in situ* CWR conservation if use of the conserved resource is not maximized.

An obvious first step in germplasm utilization is the characterization of the material. Given (1994) and FAO (1998) estimated that approximately two-thirds of globally conserved *ex situ* germplasm lacks basic passport data, 80% lacks characterization data and 95% lacks evaluation data. Given (1994) estimated that only approximately 1% of genebank accessions are appropriately documented and ready for use. The figures are unavailable for *in situ* conserved germplasm, but it seems likely that there are currently no genetic reserves where the conserved target species are fully characterized, evaluated and ready for use. In fact, it can be expected that there is still a long way to go before we can reach this point. Unless the professionals involved with CWR conservation and use can ensure that conserved germplasm is held in a form suited for breeders and other user groups and that there is a seamless gradation of conservation into utilization, the situation is unlikely to change. Pre-breeding activities are essential to facilitate the utilization of CWR in breeding programmes but unfortunately, public sector funding of such activities is quickly disappearing and genebanks are usually not equipped to carry out such critical tasks.

A major step forward in improving the accessibility of *ex situ* conserved CWR germplasm was taken by the creation of the European Central Crop Databases (http://www.ecpgr.cgiar.org/Databases/Databases.htm). These databases hold passport data and, to varying degrees, characterization and primary evaluation data of the major collections of the crops in Europe. A further step was the establishment of a European Internet Search Catalogue of *Ex Situ* PGR Accessions (EURISCO) (http://eurisco.ecpgr.org/) which is periodically updated from national inventories. These databases are an effective way of promoting the use of *ex situ* conserved CWR. At present, the data included are limited, but this is likely to improve with iteration further aiding germplasm selection by the PGR user community in Europe. The databases of *ex situ* material do not only provide a means of sourcing local material for *in situ* introduction or reintroduction by matching germplasm passport data with the conservation site details but also aid utilization of *in situ* diversity itself, when this has been safety-duplicated in genebanks. Through gap analysis of the accessions contained in the systems, a more systematic safety duplication of *in situ* genetic diversity is likely to be promoted.

An equally important step forward was taken in improving access to *in situ* conserved diversity through the development of the Crop Wild Relative Information System (CWRIS) (PGR Forum, 2005) and CWR Catalogue for Europe and the Mediterranean (Kell *et al.*, 2007). However, there is still much to be done to improve the recording and management of data associated with CWR germplasm samples conserved *ex situ* and *in situ*. In addition, the European Genebank Integrated System (AEGIS), in which participating countries identify so-called European Accessions according to agreed criteria, will form a dispersed European *ex situ* Collection for a given crop/crop gene pool. Recently, a discussion has ensued regarding the possibility of integrating *in situ* conserved genetic resources into this system which will further enhance the *in situ/ex situ* linkage and facilitate complementary gene pool conservation. Although several technical questions still need to be answered, the designation of *in situ* populations as 'European populations' would raise their visibility status at the European level and help ensure that their conservation becomes a legally binding commitment on the countries that are signatories of a Collective Memorandum of Understanding.

To improve usage of *ex situ* conserved germplasm within *in situ* conservation projects, the breadth of the species conserved *ex situ* needs to be widened. Currently, only 4% of government-funded genebanks and 14% of CGIAR genebank accessions are of wild species (FAO, 1998) – the vast majority of collections are devoted to advanced breeders material and landraces. If *in situ*

CWR conservation projects started to use *ex situ* germplasm for *in situ* population enhancement, a much broader representation of total plant diversity in *ex situ* collections would be required with maximum ecogeographic representation.

The better the quality of the sample transferred from *in situ* conservation to *ex situ*, the more likely it is to be used. Therefore, the collector must try to ensure that the sample is of a sufficient size to avoid the need for regeneration and is representative of the full range of genetic variation found in the population sampled (Brown and Marshall, 1995; Hawkes *et al.*, 2000). Similarly, the more complete the associated passport data collected, the more useful the germplasm accessions are likely to be to the end-users.

Usage of *in situ* conserved germplasm can be improved by publicizing reserve holdings and where samples may be duplicated *ex situ*. Following *ex situ* duplication, the existence of novel diversity can be signaled to potential users by publishing reports. Once the genetic reserve is established, the reserve manager can equally publish a review of the material found in the reserve, including such details as the target species' ecogeographic characteristics, and initial characterization and pre-evaluation data. For example, knowing that an accession has been sampled from a genetic reserve adjacent to the sea would be useful to the breeder trying to locate material to use in irrigated, increasingly saline soils.

Unfortunately, if the people who conserve *ex situ*, those who conserve *in situ* and those who utilize germplasm are seen as being in three distinct professions, often located in three distinct remote locations, integration of *ex situ* and *in situ* conservation with use is likely to remain slow. However, utilization can be improved by bringing conservationists and germplasm users together both physically and professionally. The mixing of these communities and the need to link conservation to use are at the foundation of the newly established CWR Specialist Group (Dulloo and Maxted, 2007).

6.6 Conclusion

Conservation of wild plant genetic diversity in genetic reserves is the only practical option for conserving the full range of their genetic diversity and is essential if we are to ensure that it continues to interactively evolve in relation to the respective ecosystems. However, the *ex situ* conservation link is fundamental to ensuring the long-term sustainability of *in situ* activities, as *ex situ* conserved germplasm offers a means of *in situ* population enhancement, provides necessary safety duplication against predictable and unpredictable environmental changes, eases *in situ* management and facilitates researchers and breeders controlled access to targeted germplasm resources for immediate utilization. The level of *ex situ* linkage and its modality, which will depend on the conservation objectives, cost–benefit decisions and local, national or international considerations, is critical to the success of *in situ* conservation. Each *in situ* conserved species is likely to require specifically targeted *ex situ* complementary measures. The result of this complementary action should be safer conservation of genetic diversity and enhanced utilization, which in turn is likely to form the basis for the sustainability of the *in situ* conservation activities themselves.

References

Adams, R.P. (1997) Conservation of DNA: DNA banking. In: Callow, J.A., Ford-Lloyd, B.V. and Newbury, H.J. (eds) *Biotechnology and Plant Genetic Resources Conservation and Use*. CAB International, Wallingford, UK, pp. 163–174.

Brown, A.D.H. and Marshall, D.R. (1995) A basic sampling strategy: theory and practice. In: Guarino, L., Ramanatha Rao, V. and Reid, R. (eds) *Collecting Plant Genetic Diversity: Technical Guidelines*. CAB International, Wallingford, UK, pp. 75–92.

de Vicente, M.C. (ed.) (2006) *DNA Banks – Providing Novel Options for Genebanks?* Topical reviews in agricultural biodiversity. International Plant Genetic Resources Institute, Rome, Italy.

Dulloo, E. and Maxted, N. (2007) The crop wild relative specialist group of the IUCN species survival commission. In: Maxted, N., Ford-Lloyd, B.V., Kell, S.P., Iriondo, J.M., Dulloo, E. and Turok, J. (eds) *Crop Wild Relative Conservation and Use*. CAB International, Wallingford, UK.

Dulloo, M.E., Ramanatha Rao, V., Engelmann, F. and Engels, J. (2005) Complementary conservation of coconuts. In: Batugal, P., Rao, V.R. and Oliver, J. (eds) *Coconut Genetic Resources*. IPGRI-APO, Serdang, Malaysia, pp. 75–90.

Engelmann, F. (1997) *In vitro* conservation methods. In: Ford-Lloyd, B.V., Newbury, J.H. and Callow, J.A. (eds) *Biotechnology and Plant Genetic Resources: Conservation and Use*. CAB International, Wallingford, UK, pp. 119–162.

Engels, J.M.M. and Visser, L. (eds) (2003) *A Guide to Effective Management of Germplasm Collections*. International Plant Genetic Resources Institute Handbooks for Genebanks 6. International Plant Genetic Resources, Rome, Italy.

Engels, J.M.M. and Wood, D. (1999) Conservation of agrobiodiversity. In: Wood, D. and Lenné, J.M. (eds) *Agrobiodiversity: Characterization, Utilization and Management*. CAB International, Wallingford, UK, pp. 355–385.

Engels, J.M.M., Dulloo, M.E. and Hoekstra, F. (2007) The role of pollen and pollinators in long-term conservation strategies of plant genetic resources. *9th International Pollination Symposium*, Iowa State University, Iowa.

FAO (1998) *State of the World's Plant Genetic Resources for Food and Agriculture*. FAO, Rome, Italy.

Given, D.R. (1994) *Principles and Practice of Plant Conservation*. Chapman & Hall, London.

Guarino, L., Ramanatha Rao, V. and Reid, R. (eds) (1995) *Collecting Plant Genetic Diversity: Technical Guidelines*. CAB International, Wallingford, UK.

Hawkes, J.G., Maxted, N. and Ford-Lloyd, B.V. (2000) *The* Ex Situ *Conservation of Plant Genetic Resources*. Kluwer, Dordrecht, The Netherlands.

Heywood, V.H. (1991) Developing a strategy for germplasm conservation in botanic gardens. In: Heywood, V.H. and Wyse-Jackson, P.S. (eds) *Tropical Botanic Gardens: Their Role in Conservation and Development*. Academic Press, London, pp. 11–23.

Hoekstra, F.A. (1995) Collecting pollen for genetic resources conservation. In: Guarino, L., Rao, V.R. and Reid, R. (eds) *Collecting Plant Genetic Diversity: Technical Guidelines*. CAB International, Wallingford, UK.

IPCC (2007) *Climate Change 2007: Fourth Assessment Report*. Intergovernmental Panel on Climate Change Secretariat, Geneva, Switzerland.

Janick, L.V., Kim, Y.H., Kitto, S. and Saranga, Y. (1993) Desiccated synthetic seed. In: Redenbaugh, K. (ed.) *Synseeds, Applications of Synthetic Seeds to Crop Improvement*. CRC Press, Boca Raton, Florida, pp. 11–33.

Kell, S.P., Knüpffer, H., Jury, S.L., Ford-Lloyd, B.V. and Maxted, N. (2007) Crops and wild relatives of the Euro-Mediterranean region: making and using a conservation catalogue. In: Maxted, N., Ford-Lloyd, B.V., Kell, S.P., Iriondo, J.M., Dulloo, E. and Turok, J. (eds)

Crop Wild Relative Conservation and Use. CAB International, Wallingford, UK.

Laliberté, B. (1997) Botanic garden seed banks/genebanks worldwide, their facilities, collections and network. *Botanic Gardens Conservation News* 2(9),18–23.

Maxted, N., Ford-Lloyd, B.V. and Hawkes, J.G. (1997) Complementary conservation strategies. In: Maxted, N., Ford-Lloyd, B.V. and Hawkes, J.G. (eds) *Plant Genetic Conservation: The* In Situ *Approach*. Chapman & Hall, London, pp. 15–39.

Osborn, T.J. and Briffa, K.R. (2005) The spatial extent of 20th-century warmth in the context of the past 1200 years. *Science* 311, 841–844.

PGR Forum (2005) *Crop Wild Relative Information System (CWRIS)*. University of Birmingham, Birmingham, UK. Available at: http://cwris.ecpgr.org/.

Rao, N.K., Hanson, J., Dulloo, M.E., Ghosh, K., Nowell, D. and Larinde, M. (2006) *Manual of Seed Handling in Genebanks*. Handbooks for Genebanks No. 8. Bioversity International, Rome, Italy.

Reed, B., Engelmann, F., Dulloo, M.E. and Engels, J.M.M. (2004) *Technical Guidelines on Management of Field and* In Vitro *Germplasm Collections*. Handbooks for Genebanks No. 7. IPGRI, Rome, Italy.

Sackville Hamilton, N.R., Engels, J.M.M., van Hintum, T., Koo, B. and Smale, M. (2002) Accession management combining or splitting accessions as a tool to improve germplasm management efficiency. *IPGRI Technical Bulletin No.5*. IPGRI, Rome, Italy. pp. 66.

Sharrock, S. and Engels, J.M.M. (1997) *Complementary Conservation*. INIBAP Annual Report 1996. INIBAP, Montpellier, France, pp. 6–9.

Simmonds, N.W. (1962) Variability in crop plants, its use and conservation. *Biological Reviews* 37, 422–465.

Thomas, C.D., Cameron, A., Green, R.E., Bakkenes, M., Beaumont, L.J., Collingham, Y.C., Erasmus, B.F.N., Ferreira De Siqeira, M., Grainger, A., Hannah, L., Hughes, L., Huntley, B., Van Jaarsveld, A.S., Midgley, G.F., Miles, L., Ortega-Huertas, M.A., Peterson, A.T., Phillips, O.L. and Williams, S.E. (2004) Extinction risk from climate change. *Nature* 427, 145–148.

Thormann, I., Dulloo, M.E. and Engels, J. (2006) Techniques for *ex situ* plant conservation. In: Henry, R.J (ed.) *Plant Conservation Genetics*. Haworth Press, New York, pp. 7–36.

Thuiller, W., Lavorel, S., Araujo, M.B., Sykes, M.T. and Prentice, I.C. (2005) Climate change threats to plant diversity in Europe. *Proceedings of the National Academy of Sciences USA* 102(23), 8245–8250.

Van Vuuren, D.P., Sala, O.E. and Pereira, H.M. (2006) The future of vascular plant diversity under four global scenarios. *Ecology and Society* 11, 25.

7 Final Considerations for the *In Situ* Conservation of Plant Genetic Diversity

J.M. Iriondo,[1] M.E. Dulloo,[2] N. Maxted,[3] E. Laguna,[4] J.M.M. Engels[2] and L. Maggioni[2]

[1]Área de Biodiversidad y Conservación, Depto. Biología y Geología, ESCET, Universidad Rey Juan Carlos, Madrid, Spain; [2]Bioversity International, Rome, Italy; [3]School of Biosciences, University of Birmingham, Edgbaston, Birmingham, UK; [4]Centro para la Investigación y Experimentación Forestal (CIEF), Generalitat Valenciana. Avda. País Valencià, Valencia, Spain

7.1 Costs and Benefits of *In Situ* Conservation: an Economic Assessment

In a world of limited resources where governments and administrations are subject to the mandates of annual budgeting and monetary goals, the economic dimension of any initiative becomes a critical point for its success. This is certainly true for current conservation actions but is likely to become the case for the *in situ* conservation of plant genetic diversity in protected areas. We may all agree on the benefits of employing such programmes, but how do we estimate the establishment and operating costs of such initiatives and how do these compare to the benefits? What models for cost–benefit analysis for genetic reserves should we use? If the benefits of investing in the conservation of plant genetic diversity cannot be demonstrated, it is unlikely that the necessary financial and human resources will be available for the proposals and techniques described in this book.

Cost–benefit analyses are management tools designed to help make decisions regarding the implementation of a particular proposal or project. Not only do they assess the cost of project implementation against the benefits accrued, but they also evaluate the costs and benefits involved in not implementing that proposal. To illustrate the procedure Primack (2006) provides the following illustration: during a feasibility study for a logging operation that would remove an area of forest, an economist might compare the income generated by the logging with the income and resources lost due to damage to game animals, medicinal plants, clean water and fish, a scenic walk through a grove of large trees, rare bird species and wildflower populations. Perhaps several scenarios are envisaged, e.g. clear-felling, partial sustainable felling and no felling, and the cost–benefit analysis provides a means of comparing the impact of the different scenarios. The time dimension is an added complication. How can we assess today which genetic diversity is essential for tomorrow's plant breeders or whether the sale price of wood pulp will increase or fall?

The initiatives in *in situ* conservation of plant genetic diversity, as well as many other proposals where the environment is involved, constitute a challenge to the application of this methodology. While the tangible income generated by habitat destruction and economic costs of conservation action can be estimated and quantified quite easily, the benefits are much more difficult to translate into monetary terms (Millennium Ecosystem Assessment, 2005). How do we estimate and quantify a scenic walk through a grove of large trees? Siikamäki and Layton (2006) analyse the cost-effectiveness of incentive payment programmes, applied to the Finnish non-productive forests, providing a good review on cost–benefit analysis including the immaterial benefits.

The specific proposals linked to the *in situ* conservation of plant genetic diversity in protected areas that we deal with in this book are a bit simpler to handle. Since these proposals are projected to take place in sites that are already protected, many of the indirect benefits that would normally be associated to the proposal, such as ecosystem services, recreation and cultural and spiritual values of the site, do not have to be taken into account initially because they are being achieved both in the proposal and non-proposal scenarios. Thus, in this case, the benefits to take into account are essentially those directly derived from the conservation and characterization of a target crop wild relative (CWR) population. Similarly, we do not have to evaluate opportunity costs, i.e. the costs in terms of an alternative opportunity for this site foregone and the benefits which could be received from it. The site is already a protected area in both scenarios, so this issue does not have to be considered. Therefore, the costs involved in the proposal are basically the implementation and operating costs and the additional benefit is the use potential of the genetic diversity conserved for use as gene donors.

Bearing in mind that legal protection status does not necessarily correspond with effective protection of the site, it is advisable to always keep track of the indirect benefits. When protection is negligible or non-existent, the consideration of indirect benefits becomes a major issue. As difficult as it may be to assign values to ecosystem services, recreation and cultural and spiritual values, it is essential to take them all into account as they may be key for shifting the decision from one alternative to the other. Chan *et al.* (2006) provide a good revision on ecosystems services, whereas Naidoo and Ricketts (2006) provide a large-scale example of cost–benefit analysis applying ecosystem services. Similarly, opportunity costs will also have to be considered when authorities have the power to reverse existing protection status or when the site has no protection at all.

This economic assessment is planned from the perspective of a protected area manager when the establishment of a genetic reserve for a particular CWR population is being considered. For the calculation of the implementation costs, we have considered that the protected areas already have a minimum staff and infrastructure (building for offices and storage of working material). Therefore, the costs involved are mainly those necessary for the identification of the target species and populations in the protected area and for the design of the management plan. At present, in many cases, the identification of the target species and populations in a protected area is likely to be achieved as a result of national or regional projects focused on the *in situ* conservation of a particular group of species of interest. To illustrate the scale of the potential costs associated with the establishment and running of a genetic reserve we provide an example in Box 7.1.

In contrast, it is certainly very difficult to give an estimate of the benefits provided by the establishment of a genetic reserve in a protected area. Much depends on the challenges that human society may need to confront with regard to the crops to which the target population is related. CWR have already made substantial contributions to improving food production through useful genes that have been transferred to new crop varieties. CWR genes have been used to improve the nutritional value of crops such as protein content and zinc content in wheat (Nevo *et al.*, 2004) and vitamin C content in tomato. Broccoli cultivars with high levels of anti-cancer glucosinolate compounds have been developed using genes obtained from wild populations of *Brassica oleracea* (Branca *et al.*, 2002). CWR have also provided resistance to pests and diseases in a wide range of crops, including rice (e.g. virus resistance from *Oryza nivara*; Vaughan and Sitch, 1991), potato (e.g. potato blight), wheat (e.g. powdery mildew and rusts), tomato (e.g. *Fusarium* wilt and nematodes), chickpea (e.g. ascochyta blight resistance from *Cicer echinospermum*; Collard *et al.*, 2003) and groundnut (e.g. root knot nematode and early leaf spot). Moreover, CWR are a gene source for increasing tolerance to abiotic stresses such as drought, soil salinity and extreme temperatures.

UNEP and the Bioversity International estimated that between 1976 and 1980, wild relatives contributed approximately US$340 million/year in yield and

Box 7.1. Genetic reserve cost case study.

As an illustration of the cost analysis of a genetic reserve, the experience resulting from the deployment of the network of micro-reserves in the Autonomous Community of Valencia in Spain is presented here. The calculations are based on a model genetic reserve of 20ha dedicated to the conservation of one CWR species and located in a protected area with a minimum infrastructure. The implementation cost including the initial identification of the target species and populations and the design of a management plan for one selected population is estimated to be around €6000. The operating cost items that were identified include: personnel, travel expenses, infrastructure materials, expendable materials, use of infrastructures and research through external contracts.

The costs estimates for personnel have been carried out considering that personnel are not specifically hired to serve a single genetic reserve, but, on the contrary, the working time is shared with conservation and management activities that are carried out in other genetic reserves or other activities of the protected area. Therefore, our estimate based on the experience of Valencian micro-reserves is that each person dedicates 1/20th of his/her time to the activities of each genetic reserve. Following these assumptions the estimated annual operating costs of a genetic reserve for the conservation of one CWR species are:

Items	Functions	Cost (€)
Manager	Management project design and implementation, censuses, monitoring, contacts with landowners and people affected by the programme, subsidy paperwork, land custody actions, attention to researchers, and writing of didactic units, updating of management plan, among others	1500
Manual operators	Control of invasive species, ploughing or vegetation removal, planting in reinforcement operations, fencing, support in monitoring activities, etc.	500
Travel	One visit to each genetic reserve per month, including meal costs	350
Small infrastructure material	Fence material, signposts and alike (renewal every 5 years)	400
Expendable material and services	Various consumable materials such as office supplies, software, sampling and monitoring material, as well as electricity and communication services	200
Use of basic infrastructure	Basic office of the manager and a vehicle. Proportional cost for an estimated life of 10 years per infrastructure, and estimating a 5% dedication to each genetic reserve	300
Research and other external service contracts	This estimate considers the implementation of similar research actions in many genetic reserves that could be offered in one single research project. As we previously noted the cost of censuses, regular monitoring, cartography of vegetation units, etc. is contemplated in the personnel section.	1000

Considering the maximum figures for each item, which would correspond to the genetic reserves demanding the most intensive management, these items add up to a total of €4250/year. However, in most cases where the genetic reserve holds a large 'healthy' population with no evident threats the management can be reduced to a single visit to the population per year. In this situation, if we maintain the research costs to characterize and increase the knowledge of the population, the operating costs can be as low as €1500.

disease resistance to the agricultural farm economy within the USA (Prescott-Allen and Prescott-Allen, 1986).

As long as the target population of one genetic reserve could facilitate the knowledge of novel traits and their transfer to its related crop, the resulting benefit would probably pay off the investment made, not just in that particular genetic reserve but in all the genetic reserves of the region or country where it occurs. Nevertheless, it is clear that the likelihood of obtaining such a type of benefit will depend to a great extent on the economic importance of the related crops and the traits present in the target populations in terms of adaptability to environmental conditions, resistance and/or tolerance to pests and pathogens and nutritional and/or medical properties. Therefore, the results of a cost–benefit analysis will favour the establishment of a genetic reserve, if the target population is carefully selected based on the importance of the related crop and the specific properties that this population can provide. Populations located near the distributional edge of the species or in habitats near the environmental range of the species are prone to have private alleles that are not present in other populations of the species and, therefore, will be more valuable.

A very particular case that easily shifts the balance in favour of the establishment of a genetic reserve is the case where the population under consideration is but one of the few existing in the whole world for that species. Most countries already show great interest and are involved in conservation actions to protect species from extinction. When this species is also a CWR, the interest in its conservation increases manifold.

Finally, the protected area manager should also take into account the additional perceived value of the protected area designated as a genetic reserve by the agronomic sector, administration and society in general, and the reserve-visiting public. In the latter case, just as botanic gardens have for decades highlighted displays of exotic crop plants to provide a spectacle for the visiting public, so can the protected area manager highlight CWR as an additional attraction for the public to view when visiting the site. Even as a specialist it is still thrilling to see the crop and CWR side by side and be amazed that the former is derived from the latter.

The cost–benefit analysis of establishing a genetic reserve in a protected area has been presented from the protected area manager's perspective. However, it is foreseeable that in most cases the decision will be evaluated at a higher level, such as the protected area network manager or executives at departments of environment or agriculture. In these cases, cost–benefit analysis is certainly more complex because the array of possible alternatives to assess is much greater.

As such, cost–benefit analysis provides an effective tool for planning conservation action and increasing the efficiency of our overall conservation efforts. Even though it is difficult to quantify the economic benefit of conserving individual CWR species, applying cost–benefit analysis is an especially useful tool for comparing prospective genetic reserve sites. For example, when deciding between two sites where the establishment and routine management costs are different and the perceived benefits are equal, there would be an obvious advantage in choosing the 'cheaper site'. Similarly, if the costs of genetic reserve establishment and routine management are similar, yet one site houses a close CWR of wheat and the other a close CWR of rye, the site with the wheat CWR would offer greater potential benefit as wheat is of greater economic value.

7.2 Policy Considerations

Any form of conservation should be set within a broader national, regional and even global policy context. Therefore, what follows is a brief overview of the major policy developments that might well impact on the *in situ* management of the genetic diversity of plants.

The CBD entered into force in December 1993 as a legally binding instrument for the conservation and sustainable use of biodiversity. It addressed the promotion of biodiversity conservation, the sustainable use of its components and the equitable sharing of the benefits arising from the use of biodiversity. The Convention has had a broad impact on the practice of conservation, for example, making the direct linkage of conservation to use, the preference for *in situ* conservation actions with *ex situ* acting largely as a back-up and the change from the previously accepted view of common heritage or ownership of natural resources to the notion of national sovereignty over the genetic resources within the states' borders. In addition, 'prior informed consent' and 'the country of origin' were two other fundamentally important concepts that shaped the thinking of the CBD. The aforementioned principles led to a situation that countries favoured bilateral accesses and benefit-sharing (ABS) arrangements. As the original philosophy of the CBD was strongly based on the conservation and utilization of 'wild' species, it was felt necessary to plead for a 'special' treatment of plant genetic resources for food and agriculture (PGRFA) as they are fundamentally different from biodiversity in general. This resulted in a request to FAO, as part of Resolution 3 of the Nairobi Final Act, to resolve the legal status of existing *ex situ* collections and to further develop the concept of Farmers' Rights. With respect to ABS, an Open-Ended Ad Hoc Working Group agreed on guiding principles in 2001 and since then negotiations are ongoing to establish a global access regime.

Within the context of the CBD the Global Strategy for Plant Conservation (GSPC) (CBD, 2002a) was adopted by the CBD at its sixth conference of the parties. It includes global targets that are to be achieved by 2010, such as '60 per cent of the world's threatened species conserved *in situ*; 60 per cent of threatened plant species in accessible *ex situ* collections . . . and 10 per cent of them included in recovery and restoration programmes', and specifically in relation to PGR, '70 per cent of the genetic diversity of crops and other major socio-economically valuable plant species conserved' (CBD, 2002a). In Europe, the European Plant Conservation Strategy (EPCS) was proposed and submitted to the CBD Subsidiary Body on Scientific, Technical and Technological Advice (SBSTTA) by Planta Europa and the Council of Europe (Anonymous, 2002). Its vision was 'a world in which wild plants are valued – now and in the future', and its goal was 'to halt the loss of wild plants diversity in Europe'. This was to be achieved by 2007, using 43 targets, including Target 17: 'Management plan for wild crop relatives initiated in at least one protected area in each of 5 or more European countries'; Target 24: '30% of wild crop relatives and other socio-economically and ethnobotanically important species stored in genebanks'; and Target 27: 'Manual with guidelines and case studies of best practice for integrated (*in situ* and *ex situ*) plant conservation programmes made available on the web' (Anonymous, 2002).

In the food and agriculture context, the International Treaty on Plant Genetic Resources for Food and Agriculture (ITPGRFA) specifically focuses on agrobiodiversity (FAO, 2001), its objectives being the 'conservation and sustainable use of plant genetic resources for food and agriculture and the fair and equitable sharing of the benefits arising out of their use'. Article 5 states that each Contracting Party shall '[s]urvey and inventory plant genetic resources for food and agriculture, taking into account the status and degree of variation in existing populations, including those that are of potential use and, as feasible, assess any threats to them.... Promote *in situ* conservation of wild crop relatives and wild plants for food production, including in protected areas'.

In June 2004 the ITPGRFA entered into force, a legally binding framework that provides for a multilateral system (MLS) of facilitated access and equitable benefit-sharing. This multilateral system is specifically designed to facilitate the conservation and sustainable use of agricultural crops for which access to genetic resources is critically important to ensure continuous improvement (Moore and Tymowski, 2005). ABS conditions only apply to the genetic resources of species or gene pools that are included in Annex I of the treaty (i.e. 35 crop gene pools, in many instances including the related wild species to the crops as well as 29 temperate forages). At the first meeting of the Governing Body a standard material transfer agreement (sMTA) was agreed upon as well as other elements, including an agreement to bring the designated germplasm maintained by Consultative Group on International Agricultural Research (CGIAR) Centres formally into the treaty as well as the recognition of the Global Crop Diversity Trust as an essential element of the funding strategy (http://www.fao.org/ag/cgrfa/gb1.htm). Only germplasm belonging to species included in Annex I that fall under the control and management of governments and is formally placed in public domain, as well as germplasm managed by CGIAR Centres, form the multilateral system's material.

Among its general obligations, the treaty 'promotes *in situ* conservation of wild crop relatives and wild plants for food production, including in protected areas, by supporting, *inter alia*, the efforts of indigenous and local communities'. Among the provisions for access to multilateral system material, the treaty says that 'the Contracting Parties agree that access to plant genetic resources for food and agriculture found in *in situ* conditions will be provided according to national legislation or, in the absence of such legislation, in accordance with such standards as may be set by the Governing Body'.

A possible interpretation of this provision is that Annex I material maintained in genetic reserves that are public domain and under the management and control of the contracting parties can be considered part of the MLS. Active designation by the governments of MLS material in the genetic reserves and the inclusion of this information in a public database is strongly recommended, as it is a prerequisite to formally place the material in the public domain. Should material from a genetic reserve fall outside the scope of the MLS, such material would have to be acquired under prevailing national access and benefit-sharing legislation of the country in question. However, it should also be kept in mind that the exercise of backing up *in situ* populations with *ex situ* accessions will, depending on national law, enable this germplasm to exit the grey area of bilateral systems and enter the MLS for access and benefit-sharing.

The current threats faced by plant genetic diversity from genetic erosion and extinction were further recognized by the CBD 2010 Biodiversity Target (CBD, 2002b), which committed the parties 'to achieve by 2010 a significant reduction of the current rate of biodiversity loss at the global, regional and national level as a contribution to poverty alleviation and to the benefit of all life on earth'. To address this target, we need to assess biodiversity change and threats, which requires precise knowledge of what biodiversity exists. Within the context of plant genetic diversity this is indeed a challenge, as our knowledge of what wild plant diversity actually exists where, let alone the genetic diversity within those wild species, remains rudimentary even for the closest wild relatives of socio-economically important species.

Many of the more recent policy initiatives are associated with well-focused, time-bound, measurable targets, rather than open-ended aims, which obviously permits a better assessment of the success or failure of previously established conservation actions, facilitates monitoring and reporting, and provides a guide for future actions. However, this focus has highlighted the need to ensure that the targets are clear and unambiguous, especially bearing in mind the difficulties of defining biodiversity in a precise and measurable manner (Heywood and Dulloo, 2006). There should also be a reasonable expectation that the goals can be met because establishing a target in itself will not enhance conservation.

7.3 Global Strategy for CWR Conservation and Use

The conservation and use of CWR involves a plethora of government agencies, NGOs, universities, commercial enterprises and other institutions at the national level, and UN agencies, CG Centres, IGOs and NGOs at regional and global levels. It is clear that there was an urgent need for some level of coordination of the activities of these various bodies, both at local and global levels (Heywood *et al.*, 2007a). Therefore, it was felt by the participants in the First International Conference on Crop Wild Relative Conservation and Use that it would be valuable to propose an international strategy that would bring the different strands together, and provide guidance for all those engaged in national, regional and global activities concerning CWR. The conference provided a platform for the development of a 'Global Strategy for CWR Conservation and Use', during which delegates debated the content of the strategy (see Heywood *et al.*, 2007b). Its international adoption is now being led by FAO who sees it becoming an integral component of the ITPGRFA.

The strategy essentially provides an action plan for nations and regions to refer to in addressing the critical issues of effective CWR conservation and use. The main objectives of the strategy (Heywood *et al.*, 2007a; Appendix 1) are:

1. Prepare national CWR strategic action plans;
2. Prepare national CWR inventories;

3. Establish a global mechanism/clearing house for CWR conservation and use;
4. Create national priority CWR lists and identify priority CWR sites;
5. Create regional and global CWR priority lists and identify priority CWR sites;
6. Establish protocols for CWR information management and dissemination and provide national and global CWR information management systems;
7. Develop effective means of conserving and using CWR *in situ*;
8. Develop effective means of conserving and using CWR *ex situ*;
9. Assess CWR conservation and threat status;
10. Ensure effective security and legislation for CWR;
11. Promote sustainable utilization of CWR;
12. Initiate education and public awareness programmes on the importance of CWR.

Within each of these broad objectives a series of practical targets are suggested, based on the existing experience and knowledge. These targets include developing national priority CWR lists for conservation action, undertaking gap analysis of national CWR representation in national *ex situ* collections, applying IUCN red list criteria to all priority CWR taxa and promoting sustainable use of CWR in breeding programmes.

Within the context of Objective 7 'Develop effective means of conserving and using CWR *in situ*' the following targets are proposed:

1. Establish globally, and within each region, a small number of priority sites (global = 100, regional = 25) for the establishment of active CWR genetic reserves. These reserves should form an interrelated network of internationally, regionally and nationally important CWR genetic reserve sites for *in situ* conservation.
2. National action to be taken to record the presence of CWR in each country's protected areas system;
3. Each country to assess whether the existing network of protected areas adequately represents the full range of national CWR diversity, and suggest additional reserve locations where required;
4. Link CWR *in situ* reserve sites with other current initiatives, such as the Important Plant Area initiative and Natura 2000 network, and where appropriate establish genetic reserves linked to these initiatives;
5. Encourage UNESCO MAB to complete its floristic inventories in MAB Reserves and highlight which CWR are known to occur in each;
6. Raise awareness among protected area managers of the importance of CWR and request they take into account the maintenance and conservation needs of CWR when drawing up or revising management plans;
7. Involve local communities in planning community conservation of CWR and encourage them to participate in the management of reserves and other protected or non-protected areas in which CWR are known to occur;
8. Examine the potential role of micro-reserves in CWR conservation;
9. Countries and agencies to review the possibilities of conservation of CWR outside protected areas, including within agroecosystems;
10. Countries and agencies to review possibilities for conservation of CWR outside protected areas via policy decisions (easements, set-aside and other appropriate mechanisms);

11. Promote traditional farming systems for both landrace and CWR conservation;
12. Establish protocols for the management and monitoring of genetic diversity in CWR populations;
13. Publish case studies for the complete genetic reserve location, establishment and routine maintenance process to act as templates for subsequent projects;
14. Publish protocols and examples of the integration of CWR *in situ* conservation and use as a means of promoting CWR use.

The implementation of the strategy is strongly recommended as a means of securing the necessary plant genetic diversity for future generations and so underpinning future food security.

The publication of this text is itself a significant step towards implementing the strategy. Once the 12 objectives have been implemented through the ITPGRFA by the target date of 2015, it is foreseen that it will revolutionize national, regional and global efforts to conserve and use CWR diversity. As CWR species are not fundamentally different from other wild plant taxa, implementation of the strategy will also benefit the conservation of non-CWR wild plant species and hopefully raise awareness in the broader conservation community of the need to conserve the maximum genetic diversity of a species.

7.4 Present Initiatives in CWR *In Situ* Conservation

Given the importance that the most recent international agreements and strategies render to CWR and *in situ* conservation, we can predict that the next decade is going to experience great activity in this field worldwide. Currently, there are a few examples of its implementation throughout the world, mainly sponsored by international organizations.

In Europe, the 3-year project, European Crop Wild Relative Diversity Assessment & Conservation Forum (PGR Forum) funded by the Fifth Framework Programme of the European Commission, has been a forum for the assessment of taxonomic and genetic diversity of European CWR and the development of appropriate conservation methodologies. It has greatly catalysed *in situ* conservation of CWR activities, collating the people and projects that were scattered throughout Europe, and coordinating and promoting initiatives in the fields of inventorying, data management and the development of methodologies and standards for the specific problems and challenges posed by this type of conservation.

The project brought together 23 partners from 21 European countries with the addition of partners representing IUCN – the World Conservation Union and Bioversity International. A broad cross section of the professional European PGR community was represented, including conservationists, taxonomists, plant breeders, information managers, policy makers and end-users. PGR Forum created the PGR Forum Crop Wild Relative Information System (CWRIS) (Kell *et al.*, 2007) providing access to European CWR data (http://www.cwris.ecpgr.org) (Kell *et al.*, 2005). CWRIS includes all socio-economically important species occurring in Europe and the Mediterranean region and their wild relatives, including food, fodder and forage, medicinal plants, condiments, ornamentals, forestry species, as well as plants used for industrial purposes, such as oils and fibres. This forum also developed methodologies for creating national and regional CWR inventories, assessing CWR threat and conservation status, undertaking CWR conservation gap analysis, establishing conservation priorities for CWR, managing CWR data, with a particular emphasis on site and population data, *in situ* genetic population management, and assessing genetic erosion and genetic pollution.

At present, the European project 'An Integrated European *In Situ* Management Workplan: Implementing Genetic Reserves and On Farm Concepts (AEGRO)' funded by Council Regulation (EC) N°870/2004 follows the path opened by PGR Forum and will establish pilot projects for the implementation of genetic reserves in wild relatives of key crops such as *Beta*, *Brassica* and *Avena.*

The possibility of including *in situ* conserved genetic resources in the European Genebank Integrated System (AEGIS) certainly deserves due attention, as it would be an opportunity to actually integrate components of the same gene pool into a duly integrated conservation approach, irrespective of whether the material is conserved *in situ* or *ex situ.* Regarding *ex situ* conserved accessions, participating countries have approved to identify the so-called Most Appropriate Accessions (MAAs) according to agreed criteria, to form the dispersed European Collection for a given crop/crop gene pool. These countries have agreed to accept long-term conservation responsibility for these accessions and to make them available to participating partners. Due to the current *ex situ* focus of AEGIS, no *in situ* conserved material can be considered until a decision is made to also include unique and important populations of crops and/or their wild relatives into the AEGIS system. In preparation for such a decision, a number of questions and issues will have to be addressed and resolved, including the question of exactly what the unit of *in situ* conserved genetic resources is, i.e. how to define a population or another subunit of a conserved taxon in operational terms? A related question is what exactly a country 'designates' to AEGIS, i.e. the described genetic diversity that grows in a properly described/defined area, only the area in geographic terms, the geographic area and defined management practices, or what? In legal terms it will require a proper understanding and definition of how to integrate material that is being conserved in a dynamic manner and that might well change from one year to another into a system that is based on genetically defined material that is expected not to change its genetic integrity over the years.

AEGIS can therefore be seen as a framework that could guarantee efficient safeguarding of unique (*ex situ* and *in situ*) genetic diversity, at the same time avoiding duplicate efforts throughout Europe (i.e. selecting the most appropriate populations for priority conservation at the regional level).

There is a long list of projects related to the *in situ* conservation of CWR, some of them concerning genetic reserves in protected areas, which have been carried out at a local or national scale in different parts of the world. A full account of these projects is outside the scope of this book; however, many of them are listed in the review of Meilleur and Hodgkin (2004) and Heywood and Dulloo (2006).

At the global scale '*In situ* conservation of CWR through enhanced information management and field application' is a UNEP/GEF-supported project that addresses national and global needs to improve global food security through effective conservation and use of CWR. This large, multifaceted, 5-year project was launched in 2004 and brings together five countries (Armenia, Bolivia, Madagascar, Sri Lanka and Uzbekistan) and six international organizations (Bioversity as the project manager, the Food and Agriculture Organization of the United Nations (FAO), Botanic Gardens Conservation International (BGCI), the United Nations Environment Programme's World Conservation Monitoring Centre (UNEP-WCMC), the World Conservation Union (IUCN) and the German Federal Agency for Agriculture and Food (BLE)). The project has four major components. The first two focus on the systematic compilation, access and use of information related to CWR, whereas the last two are dedicated to the improvement of country capacity and raising awareness about the need for the conservation of plant genetic resources (PGR) and especially CWR.

7.5 Global, National, Monographic and Site-specific Approaches to Genetic Reserve Network Establishment

Based on the experiences of these international projects and the aims of the Global Strategy for CWR Conservation and Use, it seems clear that in the medium term the focus will be on the establishment of global, regional and national networks of genetic reserves of groups of species that currently hold a greater socio-economic interest. However, there may be multiple approaches to establishing such networks and these can be characterized as global, national, monographic and site-specific (N. Maxted *et al.*, Birmingham, 2007, unpublished data).

7.5.1 Global approaches

One of the global goals in the conservation of plant genetic diversity is to ensure that the conserved sample of that diversity is maximized; the best possible representation of overall total genetic is contained *in situ* in genetic reserves or preserved using *ex situ* techniques. Within the food and agriculture context this would mean focusing on the crop gene pools, giving priority to the major crop gene pools because of their economic value. As such if a global approach to genetic reserve network establishment is being taken, a first step would be to prioritize the major crop gene pools, identify the most important CWR they contain and carry out ecogeographic and genetic diversity surveys on the target taxa. The localities of the populations can be compared against the existing protected areas over a GIS platform to help identify candidate protected areas holding populations of interest. The ecogeographic features of the protected area add further information with regard to the potential adaptation genes that some of the populations may have. This type of analysis can also help identify gaps in the network of protected areas and other potential localities that fall outside of the network that should be conserved anyway. These are actually some of the activities that will be carried out in the implementation of the above-mentioned AEGRO project.

7.5.2 National approaches

A similar approach to that used to identify the global network could be taken at the regional or national level (see Fig. 7.1), but here the starting point need not be just the major crop gene pools present in the region or country but might be the entire flora of the region or country as was the case for the recent identification of priority sites for the establishment of genetic reserves in the UK (Maxted *et al.*, 2007), Ireland (H. Fitzgerald *et al.*, Birmingham, 2007, unpublished data) and Portugal (J. Magos Brehm *et al.*, Birmingham, 2007, unpublished data). This approach is perhaps the most objective because CWR taxa target are selected from the whole wild flora rather than those that a priori are considered a priority, i.e. the major crops. Maxted *et al.* (2007) identified the seventeen 'best' sites in the UK to establish CWR genetic reserves and these contained 152 (67%) of the 226 priority UK CWR species. They also found it would require 69 genetic reserve sites to conserve all 226 priority CWR species.

There are clear differences between CWR priorities at global, regional or national levels for establishing genetic reserves. There is a limited number of globally important CWR species, possibly those associated with the top 20 crops and their gene pools, but even if a country does not contain any of these globally important CWR species they will have nationally important CWR species. It is only by adopting this multidimensional, complementary approach which involves overlapping global, regional and national networks that the full diversity of CWR can be conserved.

7.5.3 Monographic approaches

Another alternative approach to genetic reserve establishment is associated with groups of taxonomic or agrobiodiversity specialists. Both in the taxonomic and agrobiodiversity communities the scientists who work on similar taxa or crop complexes form specialist groups, e.g. International Legume Database and Information Service, IUCN Species Survival Commission Cacti and Succulents Specialist Group, Bioversity International's Tropical Fruits Network and European Cooperative Programme for Plant Genetic Resources Forages Network. Each of these groups of specialists has a conservation remit and so may have an interest in the *in situ* genetic reserve conservation of their taxon of interest. As that interest is restricted to that taxon, their approach may be regarded as monographic. Their approach is likely to involve identification of the CWR taxa within the target taxon, depending on the number of CWR taxa identified, some form of prioritization, then ecogeographic analysis and location of specific sites for the establishment of genetic reserves.

7.5.4 Site-specific approaches

The previous approaches may be designated as top-down in the sense that they start with the goal of conserving CWR and then identify the locations where CWR are concentrated to establish genetic reserves. The final approach is the reverse, where you have an existing protected area and wish to enhance its value by designating it

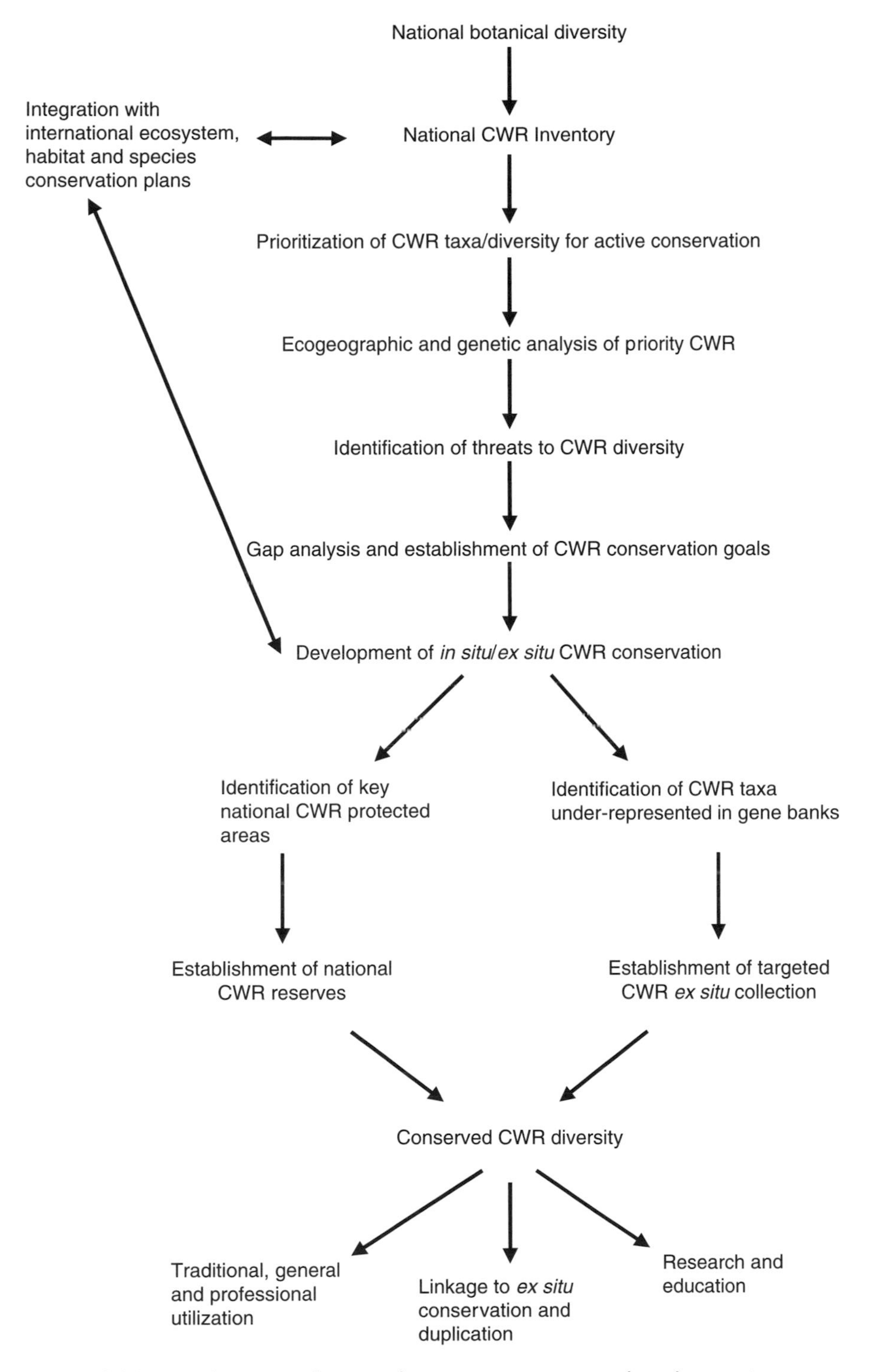

Fig. 7.1. Model for development of National CWR Strategy. (Maxted *et al.*, 2007.)

as a genetic reserve for the conservation of CWR. A reserve manager may wish to do this to increase the value of his or her protected areas and connect to a potential novel user community of plant breeders and other germplasm users.

If a National CWR Inventory exists, the individual protected area manager wishing to expand the use of his or her protected area may consult the inventory and identify the CWR present by matching with the species found in the protected area. These CWR may then be managed in an appropriate manner and so the genetic reserve established. As most protected areas are likely to have been established to conserve specific habitats or individual rare or threatened species and not explicitly to conserve CWR taxa, the highlighting of the presence of CWR species and the establishment of a genetic reserve offer an additional justification for continued funding of the reserve. The manager may also wish to publicize the presence of CWR species in the protected area to the general public as a means of emphasizing the protected area's role, for instance, in helping ensure wealth creation or food security.

7.6 Trends and Perspectives

As time goes by and several networks of this kind are developed in a country or consortia of countries, the integration of interests and the need to maximize efficiency will probably lead to multi-specific genetic reserves. For any of these projects to be successful it will be essential to count on the implication and close collaboration of plant breeders of public research centres and private companies. The focus on the species that hold the greatest socio-economic importance will be maintained in the future since the active conservation of other species will be difficult to justify following the results of a cost–benefit analysis. In this regard, Hodgkin (1997) noted that not all CWR can or should be conserved *in situ*. Many are common species whose populations are not particularly threatened, and thus priority setting within CWR species becomes a major goal that must be kept in mind.

In most countries, there is presently a great lack of coordination between the conservation activities carried out by the Department of Agriculture and those carried out by the Department of Environment. The *in situ* conservation of CWR falls entirely in an intermediate grey area and, in many instances, constitutes a no-man's-land. As this situation is progressively resolved, we may see a change of perception in protected area managers that may lead to extend the concept of genetic reserve to the whole protected area, turning the park into a great genetic reserve where major emphasis is placed on the genetic conservation of selected CWR. We have already mentioned that 80% of the species in Europe and the Mediterranean region currently fall under the CWR concept we have presented. The distinction between CWR and non-CWR species is likely to become blurrier in the future as methods to incorporate genes from virtually any living being progress. This may lead to a complete merge in the scope of CWR conservation and global biodiversity conservation. Therefore, in the long run, it may just be a matter of incorporating a new layer of genetic diversity conservation goals to the already existing goals of organism diversity and ecological diversity present in the General Management Plans of protected areas.

As humans are also part of the community and live within, or in the surroundings of, protected areas, their needs must also be taken into account. Therefore, care must be taken to encourage the participation of local communities in the process of genetic reserve conservation. Genetic reserves may provide a relevant benefit to the local community in terms of tourism and the sale of traditional products, craftwork and food that may be linked to the values of the plants being conserved (e.g. plants that may be culturally linked to people, medicinal plants, local spirits, ornamentals, etc.). Several reports have shown the successful maintenance of genetic diversity of key species through the application of traditional management principles and practices (Posey, 1984; Altieri and Merrick, 1987; Oldfield and Alcorn, 1987).

One of the biggest challenges for the medium-term/long-term will be to develop practical and suitable solutions to the problems derived by climate and global change. These solutions will inevitably have to consider the shift of genetic reserves from one place to another and the design of corridors or itineraries that can be used by more than one species. These plans will require very close collaboration among neighbouring protected areas and improved transnational coordination and cooperation. A major, essential component of these projects will be the set of species and habitat recovery techniques described in Chapter 5, which will have to be well mastered and adequately developed by then. Similarly, well-developed cooperation between germplasm banks and genetic reserves, described in Chapter 6, may prove to be vital for the success of these projects.

We have already mentioned that the network of protected areas does not and will not be able to provide a complete background for the conservation of all relevant CWR. Ingram and Williams (1993) showed that if *in situ* CWR conservation efforts were restricted to existing protected areas, many world regions rich in CWR would be left out. Therefore, additional formulas need to be developed to implement complementary networks of genetic reserves outside protected areas. This will be no easy task since we live in a world where land is becoming scarcer than ever as human population and development interests increase. This is a new field where creative alternatives for making plant genetic diversity conservation compatible with human development may emerge in the coming decades. For instance, certain 'unprotected' territories of public domain may be managed in such a way as to guarantee the conservation of target species. This could be the case of road and railway edges (Plate 5), agricultural land, hedges and riverbanks where pilot projects are already in progress in some countries (Allem, 1997; Brush, 2000; Debouck, 2000). This network of genetic reserves outside protected areas could also play a significant role in the corridors that may need to be established as a response to climate change.

7.7 Research Questions

(Contributed by Ä. Asdal, M.E. Dulloo, J.M.M. Engels, B.V. Ford-Lloyd, L. Guarino, J.M. Iriondo, A. Jarvis, S.P. Kell, H. Korpelainen, J. Labokas, E. Laguna, A. Lane and N. Maxted)

Much less is known about *in situ* genetic conservation of plant genetic diversity both inside and outside protected areas than about *ex situ* conservation of plant diversity. Hawkes (1991) concluded that *in situ* techniques were still very much in their infancy. Since then the infant has clearly matured into an adolescent, but there is still much progress to be made before we are as secure in the application of *in situ* techniques as we currently are with those applied to *ex situ* conservation. Therefore, in this final section of the text we would like to highlight what we see as some of the key research questions to be addressed in the coming years.

7.7.1 Global policy issues

The long-term sustainability of genetic reserves for plant genetic diversity conservation is largely dependent on policies relating to global environmental concerns. The impact of climate change is a case in point. More research is required to assess the regional/national impact of climate change on models for genetic reserve conservation and to make predictions of the ability of species to adapt and respond to changes in climate. One could also argue that the onset of rapid global warming could tilt the balance of genetic conservation towards the application of *ex situ* techniques because of the uncertainty about the long-term survival of *in situ* populations. However, one could also question if *ex situ* accessions held in isolation from rapid environmental changes would survive when replanted in their natural habitat.

The value of genetic diversity is in its use. In order to ensure that it is safely conserved, it is important for policy decision makers to understand the costs and benefits of genetic diversity conservation. The estimation of cost is relatively easy, but more research is warranted to determine the benefits foregone if these resources disappeared. Given that genetic *in situ* conservation is not an end in itself, how can we improve the characterization and use of *in situ* conserved plant genetic diversity by local communities, as well as by national and international user communities? The impact of novel biotechnologies on human and environmental health is a hot political issue. What are the threats of introducing transgenics into the environment and what impact would this have on *in situ* CWR conservation? Can it assist conservation or would it remove one of the central justifications for plant genetic diversity conservation? It is clear that awareness needs to be raised among the general public as well as among professional ecosystem conservationists on the value of the *in situ* conservation of genetic diversity and its impact on human well-being. It is only through this approach that we can ensure long-term policies and resource stability of genetic reserves.

7.7.2 Priorities for target taxa and genetic reserve location

We need to improve existing methodologies for genetic gap analysis and the process for prioritizing target taxa. For example, what is the best way to employ recent advances in genetic diversity assessment techniques to facilitate genetic reserve location? How much baseline data is required to make a valid decision on reserve location? And how can generalized models of genetic diversity be developed that avoid the need for extensive population sampling and genetic diversity assessment of each species? Although the selection of both target taxa and reserve sites is at least partially dependent on the remit of commissioning agencies, the limited experience available has shown these processes are often data-limited. The challenge remains of how to ensure that necessary data are available for efficient decision making. Given the likely differential impact of climate change on biodiversity hot spots, how can we incorporate predictive models of impacts into reserve placement to help ensure *in situ* sustainability?

7.7.3 Genetic reserve design

A wealth of research is available on the design of reserves focused on habitat, ecosystem and animal conservation, but much less is available on plant genetic diversity. Many research questions remain unanswered including, for example, the value of habitat corridors and stepping stones in maintaining gene flow and genetic diversity for a given species, and the role played by micro-symbionts, pollinators and other associated species in target taxon sustainability. Also, reserve design questions at the network, multi-reserve level have been less well researched. How can metapopulation theory assist in reserve design and management? What is the optimum number of populations needed to conserve the maximum amount of genetic diversity for a target taxon *in situ*?

7.7.4 Genetic reserve management

Here the research questions can be divided into generic protected area research questions and those that specifically relate to genetic diversity conservation. The former include questions such as how best to manage the eradication of plant invasives without detriment to other taxa, what specific management is required to sustain small target populations surrounded by a disturbed area precluding population immigration, how can the general public and local communities be involved in protected area conservation or even be permitted to exploit resources without detriment to the target taxa, and lastly how to frame effective legislation to protect *in situ* plant genetic diversity. As for genetic diversity conservation, we should consider that CWR are often found in pre-climax, human-disturbed habitats and continuation of certain populations is directly linked to human activities. Thus, what are the management implications for CWR management when the closely related crop is encountered locally and introgression might occur? It is also important to find out how continued agrosilvicultural activities can be integrated with target genetic conservation of plant diversity, and how we might best address and resolve conflicts of interest in conservation between different species (plant–plant or plant–animal) within the same reserve area. When attempting to conserve the genetic diversity of a range of species, which methodological approach would enable us to determine the most efficient combination of priority CWR taxa for the establishment of multi-CWR species genetic reserves?

7.7.5 Genetic conservation outside protected areas

It is assumed that *in situ* conservation is best focused in clearly delimited reserves because here the conservationist can exercise the needed control. However, this assumption should be challenged as establishing a protected area is costly and we are surrounded by many healthy plant populations that are not being deliberately managed for their continued success. So, we can ask ourselves how effective genetic conservation is outside protected areas and how it can be enhanced. What is the potential role of micro-reserves on roadsides, field margins, under orchards or in forests where the site is not designated for active long-term conservation? There is significant potential in investigating conservation synergies between plant genetic conservation and traditional, organic and biodynamic farming systems – how can we effectively combine landrace and CWR conservation in traditional agricultural landscapes?

7.7.6 Population monitoring

Here again the research questions can be divided into generic protected area and more specific genetic diversity issues. The generic questions are those related to any form of species-based conservation, such as: How can we reliably estimate minimum viable populations or minimum dynamic areas? Can we develop generalized rules that might be applied rather than adopting a species-by-species approach? A more specific genetic diversity conservation question has to do with DNA sequencing technologies and our ability to cope with the resulting information explosion. Will we be able to make effective use of this information to facilitate *in situ* reserve planning/monitoring?

7.7.7 Population and habitat recovery techniques

Restoration and habitat recovery are very challenging activities which require an understanding of community ecology in addition to the genetics of component populations and species. Techniques already exist to prioritize species that require recovery action, but how to form closer links with the restoration community so that recovery programmes for important CWR taxa are given higher priority remains a major challenge. The prospect of climate change will affect decisions on recovery actions and especially on the choice of target taxa. However, in the *in situ* plant genetic diversity conservation context, it would be interesting to investigate how target populations with limited genetic diversity might be encouraged to diversify.

7.7.8 Integration of *in situ* and *ex situ* techniques

It is generally agreed that there is a need for further integration of *in situ* and *ex situ* techniques, but what policies and scientific actions can be implemented to achieve this goal? The strengths and weaknesses of *in situ* and *ex situ* conservation have been considered both in a technical and organizational sense. Now it is critically important to work further on the management of the interface of these two approaches and related policies to ensure the sustainability and safety of the material.

Although significant steps have been taken in recent years to conserve plant genetic diversity *in situ*, only when the above-mentioned research questions are addressed and subsequent action taken, both in developed and developing countries, can we be reasonably certain that the world's PGR will be adequately preserved in nature and made available for use to benefit present and future generations. Thus, we strongly believe that the systematic *in situ* conservation and use of PGR is one of the key goals of humankind in the new millennium.

References

Allem, A. (1997) Roadside habitats: a missing link in the conservation agenda. *The Environmentalist* 17, 7–10.

Altieri, M. and Merrick, L. (1987) *In situ* conservation of crop genetic resources through maintenance of traditional farming systems. *Economic Botany* 41, 86–96.

Anonymous (2002) *European Plant Conservation Strategy*. Council of Europe and Planta Europa, London.

Branca, F., Li, G., Goyal, S. and Quiros, C.F. (2002) Survey of aliphatic glucosinolates in Sicilian wild and cultivated Brassicaceae. *Phytochemistry* 59, 717–724.

Brush, S. (ed.) (2000) *Genes in the Field: Conserving Crop Diversity on Farm*. IDRC, Ottawa, Canada and IPGRI, Rome, Italy.

Chan, K.M.A., Shaw, M.R., Cameron, D.R., Underwood, E.C. and Daily, G.C. (2006) Conservation Planning for Ecosystem Services. *PLoS Biol* 4(11). e379 doi:10.1371/journal.pbio.0040379.

Collard, B.C.Y., Pang, E.C.K., Ades, P.K. and Taylor, P.W.J. (2003) Preliminary investigation of QTLs associated with seedling resistance to ascochyta blight from *Cicer echinospermum*, a wild relative of chickpea. *Theoretical and Applied Genetics* 107, 719–729.

CBD (2002a) *Global Strategy for Plant Conservation*. Secretariat of the Convention on Biological Diversity, Montreal, Canada. Available at: http://www.biodiv.org/decisions/?lg = 0&dec = VI/9.

CBD (2002b) *2010 Biodiversity Target*. Secretariat of the Convention on Biological Diversity, Montreal, Canada. Available at: http://www.biodiv.org/2010-target/default.aspx.

Debouck, D. (2000) Perspective about *in situ* conservation of wild relatives of crops in Latin America. *In Situ* Conservation Research (Part 2). In: Vaughan, D.A. (ed.) *The 7th Ministry of Agriculture, Forestry and Fisheries (MAFF), Japan International Workshop on Genetic Resources, Proceedings*. MAFF, Tsukaba, Japan, pp. 19–39.

FAO (2001) *International Treaty on Plant Genetic Resources for Food and Agriculture*. Food and Agriculture Organization of the United Nations. Available at: http://www.fao.org/ag/cgrfa/itpgr.htm.

Hawkes, J.G. (1991) International workshop on dynamic *in situ* conservation of wild relatives of major cultivated plants: summary of final discussion and recommendations. *Israel Journal of Botany* 40, 529–536.

Heywood, V.H. and Dulloo, M.E. (2006) In Situ *Conservation of Wild Plant Species – A Critical Global Review of Good Practices*. IPGRI and FAO, Rome, Italy.

Heywood, V.H., Kell, S.P. and Maxted, N. (2007a) Towards a global strategy for the conservation and use of crop wild relatives. In: Maxted, N., Ford-Lloyd, B.V., Kell, S.P., Iriondo, J., Dulloo, E. and Turok, J. (eds) *Crop Wild Relative Conservation and Use*. CAB International, Wallingford, UK.

Heywood, V.H., Kell, S.P. and Maxted, N. (2007b) Global strategy for crop wild relative conservation and use. Available at: http://www.pgrforum.org/Documents/Conference/Global_CWR_Strategy_DRAFT_11-04-07.pdf.

Hodgkin, T. (1997) Managing the population some general considerations. In: Valdes, B., Heywood, V.H., Raimondo, F.M. and

Zohary, D. (eds) Conservation of the wild relatives of European cultivated plants. *Bocconea* 7, 197–205.

Ingram, G. and Williams, J. (1993) Gap analysis for *in situ* conservation of crop genepools: implications of the Convention on Biological Diversity. *Biodiversity Letters* 1, 141–148.

Kell, S.P., Knüpffer, H., Jury, S.L., Maxted, N. and Ford-Lloyd, B.V. (2005) *Catalogue of Crop Wild Relatives for Europe and the Mediterranean*. Available at: http://cwris.ecpgr.org/ and on CD-ROM. University of Birmingham, Birmingham, UK.

Kell, S.P., Moore, J.D., Iriondo, J.M., Scholten, M.A., Ford-Lloyd, B.V. and Maxted, N. (2007). CWRIS: a tool for managing and accessing crop wild relative information. In: Maxted, N., Ford-Lloyd, B.V., Kell, S.P., Iriondo, J., Dulloo, E. and Turok, J. (eds) *Crop Wild Relative Conservation and Use*. CAB International, Wallingford, UK.

Maxted, N., Scholten, M.A., Codd, R. and Ford-Lloyd, B.V. (2007) Creation and use of a national inventory of crop wild relatives. *Biological Conservation* 140, 142–159.

Meilleur, B.A. and Hodgkin, T. (2004) *In situ* conservation of crop wild relatives: status and trends. *Biodiversity and Conservation* 13, 663–684.

Millennium Ecosystem Assessment (2005) *Ecosystem and Human Well-being. Biodiversity Synthesis*. World Resources Institute, Washington, DC.

Moore, G. and Tymowski, W. (2005) *Explanatory Guide to the International Treaty on Plant Genetic Resources for Food and Agriculture*. IUCN Environmental Law Centre, Gland, Switzerland.

Naidoo, R. and Ricketts, T.H. (2006) Mapping the economic costs and benefits of conservation. *PLoS Biol* 4(11), e360 doi:10.1371/journal.pbio.0040360.

Nevo, E., Feldman, M., Özkan, H., Cakmak, I., Korol, A., Braun, H.J., Fahima, T., Torun, A. and Millet, E. (2004) *Triticum dicoccoides*: an important genetic resource for increasing zinc and iron concentration in modern cultivated wheat. *Soil Science and Plant Nutrition* 50(7), 1047–1054.

Oldfield, M. and Alcorn, J. (1987) Conservation of traditional agroecosystems. *Bioscience* 37, 199–208.

Prescott-Allen, R. and Prescott-Allen, C. (1986) *The First Resource: Wild Species in the North American Economy*. Yale University, New Haven, Connecticut.

Primack, R.B. (2006) *Essentials of Conservation Biology*, 4th edn. Sinauer Associates, Sunderland, Massachusetts.

Posey, D. (1984) A preliminary report on diversified management of tropical forest by the Kayapo Indians of the Brazilian Amazon. *Advances in Economic Botany* 1, 112–126.

Siikamäki, J. and Layton, D.F. (2006) Potential cost-effectiveness of incentive payment programs for biological conservation. Available at: www.rff.org/Documents/RFF-DP-06-27.pdf.

Vaughan, D.A. and Sitch, L.A. (1991) Gene flow from the jungle to farmers. *BioScience* 41, 22–28.

Index

Note: page numbers in *italics* refer to figures, tables and boxes